CYBER-PHYSICAL SYSTEMS AND INDUSTRY 4.0

Practical Applications and Security Management

CYBER-PHYSICAL SYSTEMS AND INDUSTRY 4.0

Practical Applications and Security Management

Edited by

Dinesh Goyal, PhD

Shanmugam Balamurugan, PhD

Karthikrajan Senthilnathan, PhD

Iyswarya Annapoorani, PhD

Mohammad Israr, PhD

APPLE ACADEMIC PRESS

First edition published 2022

Apple Academic Press Inc.
1265 Goldenrod Circle, NE,
Palm Bay, FL 32905 USA

4164 Lakeshore Road, Burlington,
ON, L7L 1A4 Canada

CRC Press
6000 Broken Sound Parkway NW,
Suite 300, Boca Raton, FL 33487-2742 USA

2 Park Square, Milton Park,
Abingdon, Oxon, OX14 4RN UK

© 2022 Apple Academic Press, Inc.

Apple Academic Press exclusively co-publishes with CRC Press, an imprint of Taylor & Francis Group, LLC

Library and Archives Canada Cataloguing in Publication

Title: Cyber-physical systems and industry 4.0 : practical applications and security management / edited by Dinesh Goyal, PhD, Shanmugam Balamurugan, PhD, Karthikrajan Senthilnathan, PhD, Iyswarya Annapoorani, PhD, Mohammad Israr, PhD.

Names: Goyal, Dinesh, 1976- editor. | Balamurugan, S. (Shanmugam), 1985- editor. | Senthilnathan, Karthikrajan, 1991- editor. | Annapoorani, Iyswarya, 1976- editor. | Israr, Mohammad, editor.

Description: First edition. | Includes bibliographical references and index.

Identifiers: Canadiana (print) 20210322934 | Canadiana (ebook) 20210322977 | ISBN 9781771889711 (hardcover) | ISBN 9781774639146 (softcover) | ISBN 9781003129790 (ebook)

Subjects: LCSH: Cooperating objects (Computer systems) | LCSH: Cooperating objects (Computer systems)—Industrial applications. | LCSH: Cooperating objects (Computer systems)—Security measures.

Classification: LCC TJ213 .C93 2022 | DDC 006.2/2—dc23

Library of Congress Cataloging-in-Publication Data

....

CIP data on file with US Library of Congress

....

ISBN: 978-1-77188-971-1 (hbk)
ISBN: 978-1-77463-914-6 (pbk)
ISBN: 978-1-00312-979-0 (ebk)

About the Editors

Dinesh Goyal, PhD

Dinesh Goyal, PhD, is a Professor in the Department of Computer Science and Engineering, and Director, Poornima Institute of Engineering and Technology, India. He holds a PhD from Suresh Gyan Vihar University and has 16 years of rich experience in teaching with seven years in pure R&D sector of image processing, cloud computing, and information security. He has published over 90 research papers at international publications, followed by 21 papers presented in conferences, many of which are indexed by Scopus and Web of Science. He has conducted and organized four conferences and five workshops and has five national paper publications. In addition to this, he mentored many research and doctoral scholars during his career.

Shanmugam Balamurugan, PhD

Shanmugam Balamurugan PhD, is ACM Distinguished Speaker; Founder & Chairman-Albert Einstein Engineering and Research Labs (AEER Labs); Vice Chairman-Renewable Energy Society of India (RESI), India. Formerly, he was the Director of Research and Development at Mindnotix Technologies, India. He has authored/edited 35 books, 200 international journals/conferences, and six patents to his credit. He received three post-doctoral degrees—a Doctor of Science (DSc) degree and two Doctor of Letters (D. Litt) degrees—for his significant contribution to research and development in engineering. His professional activities include roles as associate editor, editorial board member and/or reviewer for more than 100 international journals and conferences. He has been invited as chief guest/resource person/keynote plenary speaker in many reputed universities and colleges at national and international levels. His research interests include artificial intelligence, augmented reality, Internet of Things, big data analytics, cloud computing, and wearable computing. He is a life member of the ACM, ISTE, and CSI.

Karthikrajan Senthilnathan, PhD

Karthikrajan Senthilnathan, PhD, is Research Advisor and EV Charger on the Product Development Team at Revoltaxe India Pvt Ltd. He also serves as a mentor to the Atal Incubation Centre (AIC-PECF). His research interest includes power systems, back-to-back converters in power systems, cyber-physical systems, smart grids, and EV chargers. He holds two patents. Dr. Senthilnathan has authored and edited several books and has published 18 journal and conference papers. He serves as an editor for a SAGE journal, IGI Global, and Bentham eBooks. He is currently a Bentham Publisher Brand Ambassador for India. He completed his PhD at the Vellore Institute of Technology, India.

Iyswarya Annapoorani, PhD

Iyswarya Annapoorani, PhD, is Associate Professor at the School of Electrical Engineering, Vellore Institute of Technology, Chennai, India. She completed her PhD on high-voltage engineering. Her research led to a number of academic publications and presentations. Her main research interests include high voltage, back-to-back converters, smart grids, renewable energy system modeling, and power system stability and control.

Mohammad Israr, PhD

Mohammad Israr, PhD, is Professor in the Department of Mechanical Engineering at Sur University College, Sur, Sultanate of Oman. He received a Bachelor of Engineering from Mandsaur Institute of Technology, Mandsaur, affiliated to Rajiv Gandhi Proudyogiki Vishwavidyalaya, Bhopal, Madhya Pradesh (State Technological University of M.P. Accredited with "A" Grade by NACC). He received a Master of Engineering from the his Institute of Engineering and Technology, affiliated to Devi Ahilya Vishwavidyalaya, Indore, Madhya Pradesh (State University of M.P. Accredited with "A" Grade by NACC). He received PhD from Suresh Gyan Vihar University, Jaipur, Rajasthan, India (Only Private University in Rajasthan Graded "A" by NACC). His research areas include industrial engineering, operation management, logistics and supply chain.

Contents

Contributors

A. Sheik Abdullah
Thiagarajar College of Engineering, Madurai, Tamil Nadu, India

K. Iyswarya Annapoorani
School of Electrical Engineering, VIT University, Chennai, India

Arif Ansari
Marshall School of Business, University of Southern California, Los Angeles, CA 90089, USA

S. Balamurugan
ACM Distinguished Speaker
Founder & Chairman-Albert Einstein Engineering and Research Labs (AEER Labs)
Vice Chairman-Renewable Energy Society of India (RESI), India

M. K. Banga
Department of Computer Science and Engineering, Dayananda Sagar University, Bangalore, Karnataka, India

Naveen Bharathi
Software Design/Cloud Computing Consultant, Director, Navster Limited, United Kingdom

Laxmi Chand
Department of ECE, JECRC University, Jaipur 303905, India

K. Deepa
Sri Ramakrishna Engineering College, Coimbatore 600022, Tamil Nadu, India

R. Anushia Devi
School of Computing, SASTRA Deemed University, Thanjavur 613401, India

Hong-Seng Gan
Universiti Kuala Lumpur, British Malaysian Institute, Gombak, Malaysia

M. Gowtham
Department of Computer Science and Engineering, National Institute of Engineering and Institute of Technology, Mysuru, Karnataka, India

Manisha Gupta
Department of Physics, University of Rajasthan, Jaipur 302004, India

Manoj Gupta
Department of ECE, JECRC University, Jaipur 303905, India

Kamal Kant Hiran
Research Fellow, Aalborg University, Copenhagen, Denmark

K. Hari Prasath
Department of Information Technology, Vivekanandha College of Engineering for Women, Namakkal, Tamil Nadu, India

S. Jansi Rani
Sri Ramakrishna Engineering College, Coimbatore 600022, Tamil Nadu, India

Jeeva S.
Department of Data Science and Business Systems, SRM Institute of Science and Technology, Kattankulathur, Chennai, Tamil Nadu, India

J. Kanagaraj
CSIR-CLRI, Adyar Chennai 20, Tamil Nadu, India

Sai Sujith Kankipati
School of Electrical Engineering ,VIT Chennai, Tamil Nadu, India

V. Karpagam
Department of Information Technology, Sri Ramakrishna Engineering College, Coimbatore, India Coimbatore, India

A. Kavinya
Department of Information Technology, Vivekanandha College of Engineering for Women, Namakkal, Tamil Nadu, India

A. Kayalvizhi
Department of Information Technology, Sri Ramakrishna Engineering College, Coimbatore, India Coimbatore, India

Sarangam Kodati
Brilliant Institute of Engineering and Technology, Telangana, India

Arun Kumar
Department of ECE, JECRC University, Jaipur 303905, India

Sathish A. P. Kumar
Department of Computing Sciences, Coastal Carolina University, Conway, SC 29528, United States

Natarajan Meghanathan
Department of Computer Science, Jackson State University, United States

K. Murugan
Ramanujan Computing Centre, Anna University, Chennai, India

R. Parkavi
Thiagarajar College of Engineering, Madurai, Tamil Nadu, India

Sameer Patel
Department of Civil and Infrastructure Engineering, Adani Institute of Infrastructure Engineering, Ahmedabad, India

Mallanagouda Patil
Department of Computer Science and Engineering, Dayananda Sagar University, Bangalore, Karnataka, India

S. Prasidh
Director, Product Management, Bitglass, Campbell, USA

P. Priyadharshini
Thiagarajar College of Engineering, Madurai, Tamil Nadu, India

Pethuru Raj
Reliance Jio Cloud Services (JCS), Bangalore 560025, India

R. Rekha
Department of Information Technology, PSG College of Technology, Coimbatore 641004, India

Siva Sarath Chandra Reddy
School of Electrical Engineering ,VIT Chennai, Tamil Nadu, India

Ravi Samikannu
Faculty of Engineering and Technology, Botswana International University of Science and Technology, Botswana

M. K. Sandhya
Department of CSE, Meenakshi Sundararajan Engineering College, Chennai, India

N. Saranya
Sri Ramakrishna Engineering College, Coimbatore 600022, India

T. Saranya
Thiagarajar College of Engineering, Madurai, Tamil Nadu, India

N. M. Saravana Kumar
Department of Computer Science & Engineering, Vivekanandha College of Engineering for Women, Namakkal, Tamil Nadu, India

M. Senthamil Selvi
Sri Ramakrishna Engineering College, Coimbatore 600022, Tamil Nadu, India

Kanchan Sengar
Department of ECE, JECRC University, Jaipur 303905, India

Mohit Kumar Sharma
Department of ECE, JECRC University, Jaipur 303905, India

Sivabalakrishnan M.
Vellore Institute of Technology-Chennai Campus, Chennai, India

Umashankar Subramaniam
Prince Sultan University, Riyadh, Saudi Arabia

Vijayan Sumathi
School of Electrical Engineering ,VIT Chennai, Tamil Nadu, India

Suriya Praba T
School of Computing, SASTRA Deemed University, Thanjavur 613401, India

Veeramuthu Venkatesh
School of Computing, SASTRA Deemed University, Thanjavur 613401, India

A. Mohamed UvazeAhamed
Department of Computer Science, Cihan University-Erbil, Kurdistan Region, Iraq

J. Arun Venkatesh
School of Electrical Engineering, VIT University, Chennai, India

Adarsh Vidavaluru
School of Electrical Engineering ,VIT Chennai, Tamil Nadu, India

Ajay Kumar Vyas
Department of Electrical Engineering, Adani Institute of Infrastructure Engineering, Ahmedabad, India

Abbreviations

AAL	ambient assisted living
AI	artificial intelligence
AMI	advanced metering infrastructure
AVT	ambient vibration testing
BASN	body area sensor network
BF	beamforming
BTS	base transceiver station
CCEF	commutative cipher based en-route filtering scheme
CE	constant-envelope
CNN	convolutional neural network
CPES	cyber-physical energy systems
CPS	cyber-physical system
D2C	device-to-cloud
D2D	device-to-device
DAA	data aggregation and authentication
DDoS	distributed denial of service attacks
DFT	discrete Fourier transform
DL	deep learning
DNN	deep neural network
DoS	denial-of-service
DSRC	dedicated short-range communication
ECG	electrocardiogram
EHR	electronic health record
EMS	smart energy management system
EPRI	electrical power research institute
EV	electric vehicle
FDI	field device integration
FDR	full duplex relaying
FVT	forced vibration testing
GAN	generative adversarial network
GMM	Gaussian mixture model
GPU	graphics processing units
FAR	false acceptance rate

FFT	fast Fourier transform
FRR	false rejection rate
HIPAA	Health Insurance Portability and Accountability Act
HTER	half total error rate
IADS	insider attacker detection scheme
ICT	information and communications technology
IDS	intrusion detection system
IOE	internet of energy
IoT	internet of things
IPI	inter pulse interval
LSCF	least-squares complex frequency
MAC	multiple access channel
MAS	multi agent system
MD	medical devices
MIMO	multi-input-multi-output
MMSE	minimum mean square error
MOG	mixture of Gaussian
NIST	National Institute of Standards and Technology
OA	outlier analysis
ODCS	outlier detection and countermeasures scheme
OFDM	orthogonal frequency division multiplexing
PAPR	peak normal force proportion
PBAS	pixel based adaptive segmenter
PCA	principle component analysis
PMU	phasor measurement unit
QoS	quality of service
RCS	resilience control system
RFID	radio frequency identification
ROC	receiver operating characteristic
RPCA	robust principle component analysis
RPM	remote patient monitoring
SG	smart grid
SAMCON	SAMple CONsense
SAT	secure aggregation tree
SDAP	secure hop-by-hop data aggregation protocol
SEF	statistical en-route detection and filtering scheme
SGEMS	smart grid energy management system
SOBS	self-organizing background subtraction

SoC	system on chip
SSDLC	secure software development life cycle
SSI	subspace system identification
SVM	support vector machine
UAV	unmanned aerial vehicle
V2R	vehicle-to-roadside
VANET	vehicular *ad hoc* network
ViBE	visual background extractor method
VLC	visual light communication
WAMC	wide area monitoring and control
WAN	wide area network
WAVE	wireless access for vehicular environment
WBASN	wireless body area sensor networks
WSNs	wireless sensor networks
ZF	zero forcing

Preface

The growth of information and communication technologies (ICT) in the industrial growth results in Industry 4.0 with cyber-physical systems (CPS). The major factors influencing Industry 4.0 are interpretability, information transparency, and decentralized decisions.

1. **Interpretability:** The capability of physical systems and humans to connect and communicate with each other through communication protocols.
2. **Information transparency:** The capability of cyber systems to build a cybernetic copy of the physical system (cyber twin) with the enhancement of sensor data. The information requires for processing from sensor data to higher context data.
3. **Technical assistance:** Two phases of technical assistance exists in Industry 4.0:
   a. Initially, the capability of a system to support humans by comprehensively collecting the data for decision-making and rigorous fault clearance in the physical systems.
   b. Then the ability of the system is analyzed by creating faults in cyber-physical systems to identify the human interaction.
4. **Decentralized decisions:** CPS has the ability to make a verdict autonomously and on its own. In case of exceptions, conflicts or interferences, the tasks are decided at higher level.

The definition of cyber-physical systems (CPS) is the integration of physical process with embedded computation, controller, and network monitoring along with the feedback loop from physical systems. In other words CPS is given by 3C's,

- Computations
- Communications
- Control

The architecture of cyber-physical systems should be universal and/or an integration of models such as:

1. **Ambient intelligence:** The embedded system is sensitive and responsive to the physical systems. In the cyber-physical environment, the physical devices, sensors, and actuators work together with humans (man to machine and vice versa) using communication protocol.
2. **Semantic control laws:** Control law should work as occurrence–state–exploit-type law, and it practices the core of CPS control unit.
3. **Networking techniques:** Wired/wireless networks with secured connection to protect system from cyber-attacks.
4. **Event driven:** The data from the sensors are recorded as events and actions are carried out by actuators. The data are the abstraction of physical device collected by the CPS units
5. **Quantified confidence:** The data from the sensors of physical device hold the raw data for processing.
6. **Confidence:** The data/information should be confidential and protected from cyber-attacks.
7. **Digital signature and authentication code:** The data/information from the CPS should be authenticated by the publisher.
8. **Criticalness:** This specifies the critical perseverance of each event/information from the sensor data from physical device.

The advantages of implementing CPS to the physical device are:

1. **Interaction between human and systems:** For decision-making, the observing changes in physical device and fixing the boundary level is critical. CPS is required to analyze such complex systems. CPS has a two-way communication between the target and users (man to machine and vice versa).
2. **Better system performance:** CPS has the capability to provide dynamic response by feedback and reconfiguration for the sensor data and cyber infrastructure. CPS ensures the better computation of data with multiple sensors and communication devices.
3. **Faster response time:** Due to presence of fast communication capability of sensors and cyber infrastructure, it enables the dynamic control of physical device for proper utilization of collected resources from the physical device.
4. **Uncertainty:** It enables the promising behavior due to high degree of inter connectivity for a large-scale CPS coupling.

5. **Scalability:** CPS has scalability properties based on demand, and users can acquire additional infrastructure with existing cloud computing. It combines physical dynamics of the target with computational models. The communication infrastructure with software model is combined in cyber domain. The sensor data with electrical, mechanical, biological, and human comprise the physical domain.

6. **Certainty:** It ensures the CPS design is valid and trustworthy. CPS has the capability of validating the system behavior of an unknown system.

7. **Capability:** CPS allows the user to add the additional capabilities to the complex physical system.

8. **Computing and communication with physical processes:** CPS has an efficient and safest computing and communication system that reduces the need of a separate operating system for CPS.

APPLICATIONS OF CPS

Cyber-physical systems have a wide range of applications in industrial control systems (ICS), smart grid systems, medical devices, and smart cars. The methodology and case studies of CPS are discussed below.

1. **Industrial Control Systems:** ICS denotes a control systems used to enhance the control, monitoring, and production in different industries such as the nuclear plants, water and sewage systems, and irrigation systems. Sometimes ICS is called SCADA, or distributed control systems. In ICS, different controllers with different capabilities collaborate to achieve numerous expected goals. A general controller is the programmable logic controller (PLC), which is a microprocessor intended to operate uninterruptedly in unreceptive situations. This device is associated to the physical world through sensors and actuators. Usually, it is equipped with wireless and wired communication capacity that is configured depending on the surrounding environments. It can also be connected to PC systems in a control center that monitor and control the operations. An example of ICS is controlling the temperature in a chemical plant. The objective is to preserve the temperature within a definite range. If the temperature goes beyond a definite threshold, the PLC

is alerted via a wireless sensor attached to the tank, which in turn, notifies the control center of the undesired temperature change.

On the other hand, in closed-loop settings, the PLC could turn the cooling system on to reduce that tank's temperature within the desired range. The cyber interactions of the PLC ensure that there is no direct communication with physical mechanisms, such as cooling fans or the tank. This involves laptops that can directly link PLCs, communications with higher-level environments such as the control center and other remote entities, and the PLC's wireless interface that could be based on long- or short-range frequencies. In addition, cyber-physical aspects are those that connect cyber and physical aspects. The PLC, the actuator, and the sensor are all cyber-physical aspects due to their direct interactions with the physical world. The wireless capabilities of the actuator and the sensor are also considered cyber-physical. Finally, the physical aspects are the physical objects that need monitoring and control, i.e., the cooling fans and the tank's temperature.

2. **Smart Grid Systems:** The smart grid is proposed as the next generation of the power grid that has been used for eras for power generation, transmission, and distribution. The smart grid delivers numerous benefits and advanced functionalities. At the national level, it provides improved emission control, worldwide load balancing, smart generation, and energy savings. While at the local level, it permits residential consumers better control over their energy use that would be helpful economically and ecologically. The smart grid comprises of two major components: power application and supporting infrastructure. The power application is where the essential job of the smart grid is delivered, i.e., electricity generation, transmission, and distribution, whereas the supporting infrastructure is the smart component that is concerned with control and monitoring the actions of the smart grid using a set of software, hardware, and communication networks. A smart meter is connected to every residence to deliver utility companies with more accurate electricity consumption data and customers with a convenient way to track their usage data. A smart meter interfaces a household's appliances and home energy management systems on the one hand, and interfaces with data accumulators on the other. Wireless communications are the most common means to interconnect with accumulators, even though wired

communications, such as power line communications, are also available. A meter is equipped with a diagnostics port that relies on the short-range wireless interface for convenient access by digital meter readers and diagnostics tools.

The smart meter sends the measurements to an accumulator that aggregates all meters statistics in a nominated region. The accumulator sends the combined data to a distribution control center managed by the utility company. In particular, the data is sent to the AMI headend server that stores the meters data and shares it with the meter data management system that manages the data with other systems such as demand response systems, historians, and billing systems. The headend can attach/detach facilities by remotely directing commands to the meters. This feature is a double-edged sword such that it is a very efficient way to control services, yet it could be exploited to launch large-scale blackouts by remotely controlling a large number of smart meters. Cyber aspects appear in the control center where smart meters' data is stored, shared, and analyzed and based on that some decisions can be made based on the analysis. The control center can also have a cyber-physical aspect when attach/detach commands are sent by the AMI headend to smart meters. Moreover, the cyber-physical aspect is also deceptive in the smart meter itself due to its ability to perform cyber operations, such as sending measurements to utility, and physical operations, such as connecting/disconnecting electricity services.

3. **Medical Devices:** Advances in medical devices has been achieved by integrating cyber-physical systems to provide better health care services. Medical devices with cyber skills that have a physical impact on patients are gaining more interest. Such medical devices are implanted inside the patient's body, called IMDs, or worn by patients, called wearable devices. They are provided with wireless capabilities to allow communication with other devices such as the programmer, which is needed for updating and reconfiguring the devices. Wearable devices communicate with each other or with other devices, such as a remote physician or smartphone. The insulin pump and the implantable cardioverter defibrillator (ICD) are examples of the medical devices with cyber capabilities. The insulin pump can automatically or manually inject insulin injections for diabetes patients when needed, whereas the ICS is used

to sense speedy heartbeat and react by sending an electric shock to sustain a standard heartbeat rate. The insulin pump requires the continuous glucose monitor (CGM), to receive blood sugar measurements. The devices, the insulin pump, and the CGM, require small syringes to be injected into a patient's body. The insulin pump gets measurements of glucose levels from the CGM. Based on the measurements, the pump decides whether the patient needs an insulin dose or not.

The CGM sends the measurements over wireless signals to the insulin pump or other devices, such as a remote controller or computer. Moreover, some insulin pumps can be directed by a remote controller held by a patient or physician. The cyber-physical aspects, alternatively, are present in those devices that directly interact with patients implanted devices. An IMD links to the hospital by transferring measurements over an in-home router. To reconfigure an ICD, a physical nearness is required to be able to do so using a device called the programmer.

4. **Smart Cars:** Smart cars also known as intelligent cars that are friendlier to the environment, with improved fuel efficiency, safe, and have greater fun and accessibility structures. These developments are achieved by the support of a group of computers interacted together, called electronic control units (ECUs). ECUs are in control of monitoring and control of various functions such as engine emission control, brake control, entertainment and luxury features. Subject to the task to be performed, each ECU is attached to the corresponding network through a sub-network.

Each ECU in different networks communicates with each other thorough separate gateway such as CAN bus. ECUs do not have any interactions with physical components of the cars like the telematics control unit (TCU) and the media player. The TCU has a wireless interface that permits advanced functionalities such as remote software updates by car manufacturers, phone pairing, hands-free usage of phones. The cyber-physical annotations are for ECUs that can legitimately interact with physical components and manipulate them, such as the parking assist and the remote keyless entry (RKE) systems. The RKE receives signals to make a physical impact on the car by locking/unlocking doors.

CHAPTER 1

Energy Management System in Smart Grids: A Cyber-Physical System Approach

J. ARUN VENKATESH[1], K. IYSWARYA ANNAPOORANI[1*], and
RAVI SAMIKANNU[2]

[1]School of Electrical Engineering, VIT University, Chennai, India

[2]Faculty of Engineering and Technology, Botswana International
University of Science and Technology, Botswana

*Corresponding author. E-mail: iyswarya.annapoorani@vit.ac.in

ABSTRACT

Smart grids are electric grids that utilize advanced technologies of monitoring, control, and communication to supply reliable and secure energy, improve the efficiency of the system, and provide flexibility to the prosumers. The beginning of the smart grid era and the development in modern infrastructures of metering, communication, and energy storage have revolutionized the power grid. Smart grids are developed with complex physical networks and cyber systems thus enabling smart grids for the Internet of Energy (IoE). IoE is the cloud where sources of power generation and loads of power consumption are embedded with intelligence. Modern electric power grids are integrated with sensors used to provide measurements. The sensor measurements and the complex applications of various sensors cause a need for Cyber-Physical System (CPS). CPS is a class of systems that integrates physical process, computation, and networking. The CPS model of the smart grid helps in modeling and simulation for evaluation of the system performance and characteristics. The CPS model of the smart grid must enable the smart grid to be robust, allow future extensions, and compatible with web service technologies. The power generation sources and loads such as smart buildings are the physical layers, the sensors used for measurements form the cyber-physical integration and the data storage and processing using IOE forms the cyber layer of the smart grid. The CPS model of the smart grid helps in the integration of intelligent devices and allied information and communication technologies for robust and reliable operation in smart grids. In the smart grid paradigm, the energy management system has a vital role to increase the efficiency and reliability of the system. This chapter presents a CPS model for smart grid and the challenges associated with the development of the CPS model. In addition, this chapter describes the energy management system model of the smart grid.

1.1 INTRODUCTION

The complex interactive network was the first research carried by the Electrical Power Research Institute (EPRI) in 1998 for developing a complete automated reliable grid, which is the first prototype of smart grid.[1] The smart grid concept was widely accepted for the future development of power network after the proposal of Intelli-Grid in 2002.[2,3] In 2005, the European Smart-Grids Technology Platform was founded which launched a report with concepts and framework for the European smart grid in 2006.[4] This report was further modified by the U.S. Department of Energy to support a reliable and sustainable energy supply in the report "The Smart Grid" in 2007. The main challenge for many countries is to develop a smart city with socio-economic and environmental benefits, improving electricity usage in smarter way to conserve energy. There is an increase in the demand for electric supply and load patterns becoming complex in nature in recent years challenging the power grid network. To address these challenges, power engineers and researchers have proposed the concept of the CPS approach for the power system network. The combination of cyber-physical system and power system network together known as Cyber-Physical Energy Systems (CPES) is based on the measurements from the sensors through which the decision to execute the control in the distributed network is achieved. The major challenge in CPES occurs in the integration of cyber and physical components in the system. In the CPES for a unified operation of cyber and physical layer components, all events happening and decisions taken must be communicated between cyber and physical layer components, thereby increasing the capability of the system to address the issues. The development of smart sensors and their integration in the electric grid ensures the availability of original and dependable data at the control centers. The original trustworthy data help in increase of accuracy to solve problems and apply control for various applications. In the present electric grid, more amount of renewable energy is penetrated, which is not been reliably handled due to the intermittency in renewable energy availability. This challenge leads to the concept of a smart grid with improved infrastructures for communication and doing various computations in the conventional grid. The present applications of the smart grid are without an underlying framework which results in isolation and are difficult to integrate and future expansion. With the help of a reference model of CPES, the smart grid can be developed in a better way. The CPES reference model must handle large scale and long-term scenario of smart grid. The CPES model must be a generic model that describes the characteristics of the smart grid scenario with technologies and standards for smart grids. The main aim of power engineers is to design an efficient algorithm which is capable of running in real time in the grid. Another challenge to the power engineers is the unprecedented volume of data measured using the Phasor Measurement Unit (PMU) which has to be aggregated and processed based on the requirement. The coordination between the decentralized resources is the major part in the real-time operation of the grid. The communication network used in the grid must be advanced to handle the coordinated operation of the grid. The infrastructure comprises of communication network and middleware comprising the software for processing data and deployment of control. The structure of the traditional power network is shown in Figure 1.1. In this chapter, a survey of research in cyber-physical energy systems is presented with an overview of CPS. The aim of this chapter is to help power engineers and researchers to gain insights into the CPS approach to power grids.

1.2 SMART GRID

Smart grid (SG) is a widely used term with various definitions. The definition of SG is to integrate and enable Information and Communications Technology (ICT) and advanced technology with power network to make the power system efficient, economical, and sustainable. In the United States, SG refers to transforming the electric industry from centralized, producer-controlled network to consumer interactive. In Europe, SG refers to the participation and integration of all societies in European countries. In China, SG refers to a physical network based approach to ensure security, reliability, and sustainability. In IEEE Grid Vision 2050, the requirement of SG is to operate and control the entire power system comprising of all present and future technologies.

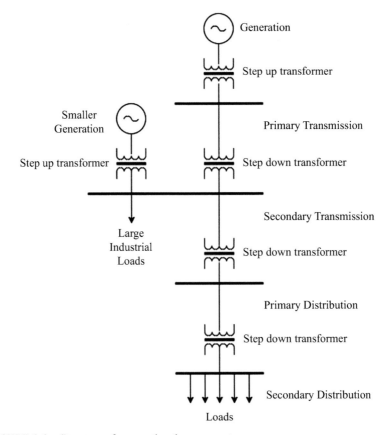

FIGURE 1.1   Structure of conventional power system.

The need of flexible, portable, safe, and secure power supply use through SG demands a reconsidering in the interaction between physical power system, the cyber systems, and users. The challenges involved in SGs are the intermittency in a renewable generation which affects power quality and stability of the system. In power demand peak, average demand plays a vital role, reduction of peak demand increases the capacity of supply without the addition of new generation units. In SGs due to the utilization of distributed generation, losses in power can be decreased by avoiding the long-distance transmission lines. The usage of smart meters, advanced sensors, and ICT helps in improving the efficiency of SGs. To achieve all the above advantages, the SGs must have the following features:

- Distributed control
- Prediction of load in advance
- Forecasting of renewable generation
- Reduction of peak demand
- Energy storage system

The solution for these issues and challenges is the CPS paradigm, which uses a systematic way to solve the issues and challenges.

1.3 CYBER-PHYSICAL SYSTEM

The U.S. National Science Foundation coined the term CPS in 2006 to describe a complex, multidisciplinary, next-generation system integrating embedded technologies in the physical world. In the United States, CPS is the integration of embedded systems and physical components, while in Europe it is the communication between cloud and human, while in china it focuses on intelligence in sensing, processing, and control. The progress in CPS is significant in the last few years and has miles to achieve its complete potential. The development is quick in sensing, analyzing, synthesizing, modeling, and control in fields of engineering and science. CPSs bring engineering and computer science together to deal with the issues and challenges. The technological challenges in bringing the two fields together are:

- Design: To achieve continuous integration, communication, and computation design is a vital infrastructure. Standard architectures and design tools are required to support the system needs. Architectures and techniques should ensure confidentiality, integrity, data availability, and protection of assets.
- Science and engineering: Integration of cyber and physical components requires fast sensing, faster processing and quick control and has to be accurate. Fast and efficient processing of large volumes of data must be present to make decisions and control actions. The

centralized control which is used traditionally does not have the speed and hence distributed control is required. Data sensing, data processing and control are the main factors involved.

1.3.1   GENERALIZED CPS MODEL

The physical layer consists of the devices which are needed to monitor and control through CPS. The physical layer consists of devices such as generators, transformers, loads, and measuring devices. The multi-input-multi-output (MIMO) CPS model is represented as,

$$\dot{x}(t) = Ax(t) + Bu(t) \tag{1.1}$$

$$y(t) = Cx(t) \tag{1.2}$$

where, A is the state matrix, B is the input matrix, C is the output matrix, x(t) is the state vector, u(t) is the input vector, y(t) is the output vector.

The control is achieved using input vector which is given by,

$$u(t) = Kx(t) \tag{1.3}$$

where, K is the connection between cyber layer control and physical layer sensors.

1.4   SMART GRID—A CPS APPROACH

SG is the integration of physical components of the power grid network and the cyber layer to achieve the characteristics of CPS. In SG, real and virtual systems are integrated where events in physical systems are communicated as input to CPS control centers and simulated to analyze the performance of the physical system. The dynamic cooperation between physical and cyber systems is achieved through communication channels. The parallel computation and distributed data help to make the decisions through CPS layers. The CPS will adapt, organize, and learn by itself and hence, it can respond for fault, attack, and emergency in SG making SG to be secure and reliable. The challenges in the CPS aspect of the SG are system is time-critical, components work together to achieve stability, regulation of voltage and frequency, and fast response subject to uncertainties and disturbances. In SG, CPS is used to reduce redundancy and improve the stability of SG. The main functions of CPS in SG are:

- Dependability
- Reliability
- Predictability
- Sustainability
- Security
- Interoperability

Many researches are carried to address the issues associated to SG and CPS. The integration of SG and CPS is known as CPES.

Architecture: The physical component of the power system network requires safety and reliability and differs from other object-oriented software. The CPS architecture must be specific for interfacing cyber and physical components to allow the system to operate in uncertainty and unpredictable conditions. The software-based components work reliably in these circumstances. The software platforms do not have timing properties and hence rethinking of the computer architectures considering the power systems characteristics is required. It needs some standards and frameworks in which physical, communication, computation components interface are based on standards of their own and interfaced together.

Communication: Communication technology is essential for efficient and effective interaction between physical layer and cyber layer components. Space and time are the aspects in communication which refers to distance and time needed for data exchange which has to be considered. The various levels of communication depend on the network size as home area network, neighborhood area network, metropolitan area network, and wide area network. Key factors which have an impact in real-time are time delay, error pockets, and delays in queues.

Modeling and Simulation: The modeling tool must be capable of supporting the network specifications, interoperability, hybrid modeling, and operation in a large scale. In the future, the SG can be large scale or small scale with distributed sources and energy demands and must be operated reliably and in a user-friendly market, involving risk analysis, risk management, security, uncertainty analysis, and co-ordination.

Cyber Security: CPS must be safe and any random failure or attacks would have harmful effects on the system. The use of cyber components such as PMU, advanced metering infrastructure (AMI) makes the system prone to attack. The SG has to be developed to detect and mitigate attacks in cyber components using, intrusion detection system (IDS). There are two types of IDS namely host-based IDS and network-based IDS.

Distributed Computation: In SG, large number of smart meters and sensors are placed at various levels, which have to process large data in sequential order. In power networks fault diagnosis, power control and reconfiguration, management and restoration are based on time and create a challenge in SG. The solution for these challenges is data mining techniques which are suitable to deal with large volumes of data. The modern computational methods grid and cloud computing platforms are used in SG to perform sub computation and local computations.

Distributed Intelligence: The large scale computation in the SGs is done using multi-agent system (MAS). An agent is an entity to control the components, capable of communicating and interacting with each other toward local/global goals. In MAS group of agents are utilized in a distributed network focusing on various applications. In SG, automation has to be in both micro and macro operational levels for decision making based on requirements.

Distributed Optimization: SG depends on global optimization and local control, where global optimization has many objectives and local control has one objective. Centralized optimization is not suitable for SG and hence distributed optimization strategies are required. MAS is used to achieve global coordination in the integration of global optimization and local control.

Distributed Control: In SG as number of components increase, the system becomes complex as many levels of control and hierarchy are present in the system. The control objectives are multiobjective with global and local requirements which can vary depending upon the operating states. The control has to produce data from physical components to analyze and control the components in the system. In SG, the control is based on physical layer, cyber layer and planning and operations layer. Figure 1.2 shows the three layers of the CPS aspect of SG.

1.5   OVERVIEW OF CPES

The power system analysis in modern systems is done with the help of computer models and is an active research topic. When computers were introduced for the power system grid, new software was developed to model the transmission and distribution system. This system was modified to compute more complex networks and compute faster by developing new software. The research in the development of new software helped in developing a distributed model of power network, parallel computation,

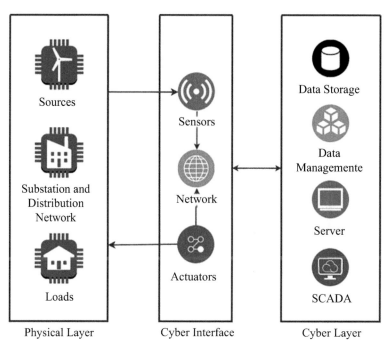

FIGURE 1.2 Cyber, physical, and interface layers of SG.

and analysis of the system. In the existing power system simulators, some are used exclusively for transmission system such as Siemens PSSE and for distribution system such as GridLab-D. Recently, active research is on the area of co-simulation. Many cloud-based software are used to model the network and perform simulation studies. In co-simulation, continuous system and discrete events simulation have been integrated to simulate and analyze the behavior of CPES. Various methods have been used to integrate various systems such as common information model. These methodologies are used for the energy management in distributed systems and various subsystems. Many researchers have utilized the CPS for the design of the power grid to analyze the reliability and security of the power system.[5-15] The research focuses on the challenges in modeling, design, and simulation of CPS. CPS is used in wide applications such as management, smart buildings cloud computation, surveillance, scheduling, monitoring, and vehicle systems.[16] In the power grid, any outage or blackouts in the power network cause a great impact on the economy and society making the operation of the power network to be critical.[17]

The CPES model is a framework of interlinks designed to achieve communication among the stakeholders. This model is not a standard for CPES but has the information regarding various technologies and standards such as the National Institute of Standards and Technology (NIST) and IEC for the smart grid. This model can be used to develop new technologies, standards, or employ new algorithms and test the performance of smart grids. Using the CPES model future standards and technology can be evolved from existing ones.

Some reference models that has been developed and discussed in literature are:

- Open Systems Interconnection Reference Model.
- Agent Systems Reference Model.
- National Institute of Standards and Technology Reference Model.
- Task-based Reference Model.

1.6   REQUIREMENTS AND CHALLENGES IN MODELING OF SMART GRID

In recent years due to the development of technologies and the use of distributed energy sources caused many challenges in monitoring and control of the power networks. Many PMUs have been placed to measure real-time data and communicate it to the control center. PMUs measure data with a high sampling rate. Wide Area Network (WAN) is used to meet with this high sample rate to create Wide Area Monitoring and Control (WAMC). WAMC can be used in power grids for many applications such as state estimation, contingency analysis, optimal power flow analysis, economic dispatch, and automatic generation control. These data collected are utilized to run the systems with control algorithms but all the data must be synchronously measured to avoid errors, which are done in the underlying framework. Thus, underlying infrastructure is an important layer for the power system applications. The applications can be functional and nonfunctional. The functional applications are the synchronization and coordination of data flow between distributed resources in the network. The nonfunctional applications are scalability (support for a large amount of PMUs and communication network), latency and predictability (time sensitive) and reconfigurability (addition or removal of components, nodes, or modifications in control algorithms).

1.7   CONCEPT OF SMART GRID ENERGY MANAGEMENT SYSTEM (SGEMS)

In the SG paradigm, the AMI devices are used for two-way communication between the utility and consumer providing opportunity for demand management by shifting the peak loads. It is an optimal management system used to provide service to monitor and manage power generation, consumption, and storage in the SGs. The communication infrastructure in the network is used to collect the information of load demand, generation, and forecasting data from all sensors to provide remote monitoring and control for various operating modes and is monitored in the control centers. The SGEMS not only provides optimal utilization of generation, but also energy storage and management functions for the system.

1.7.1   ARCHITECTURE OF SGEMS

The SGEMS center has a central controller to deliver the utility and consumer with monitoring and control functions depending on the communication. The smart meter acts as an interlinking communication between utility and consumer. The smart meters collect the data and send it to control center, which receives the control signal to optimize the demand management based on generation. Electric vehicle (EV) consumes power from SG, and also provides power back to SG in case of emergency and acts as energy storage. The distributed generations in the SG are integrated to achieve generation management and hence SG need not rely on power from central grid. Since renewable generation is intermittent in nature, energy storage system has a major role to maintain power quality, efficiency, and reliability.

1.7.2   FUNCTIONS OF SGEMS

The SGEMS must be flexible to manage and control the SG to participate in market with energy savings and load demand being met. The control services are available for the utility and consumers and they can choose the services and preferences using human–machine interface. The major functions of the SGEMS and its description are:

- Monitoring
  - Offers access to data on energy generation and demand.
  - Provides display of operational mode and status.
- Logging
  - Collect and save the data on generation from DERs, demand from loads and energy storage system.
- Control
  - The two types of control, direct implemented on equipment and control and remote control where customers monitor the load patterns and control.
- Alarm
  - Alarm is generated at SGEMS center with data on abnormalities detected in the system.
- Management
  - Management enhances the optimization and efficient utilization of energy usage in SG. It provides services such as DER management, storage management.

1.7.3   SGEMS INFRASTRUCTURES

The SGEMS infrastructure consists of a smart control center, smart meter, communication and networking system, energy storage, distributed generation, and other smart devices. With these infrastructures, SGEMS can access, monitor, control and optimize the performance of various distributed generations, loads, and other devices. SGEMS supports the integration of loads and generation with two-way communication.

1.7.3.1   COMMUNICATION NETWORK

The SGEMS have been designed based on the communication with hardware such as powerline communication and human–machine interface. Researchers work on the topic of new communication networks for the WAN. The communication network of SGEMS must meet the standards IEEE 802.15.4 for the WAN. The facility of embedding Bluetooth technology in communication can also be used in SGEMS. In SGEMS, the main components are processor for applications, communication, user, sensor, and load interface to achieve operation on the system.

1.7.3.2 SMART METER

Smart meters are used to measure the energy consumption and generation of consumer and power generation and use two-way communication to transfer data to control center and receive signals from the control center. The main functions of smart meters are measuring energy usage, two-way communication, sending data and receiving instructions, smart load shedding transition in case of failure and collection of data.

1.7.3.3 SMART ENERGY MANAGEMENT SYSTEM (EMS) CENTER

Smart **Energy Management System** (EMS) center is the brain of the entire smart grid and implements the energy management system in SG. The main functions of the smart EMS center are: receiving sent by smart meters and control panels, automated demand response, human–machine interface, online monitoring, scalability, integrating distributed resources and energy storage, forecasting renewable generations, and optimal control.

1.7.4 DER IN SMART GRIDS

The utilization of energy from renewable sources started to increase rapidly since the 1990s in various areas such as industries, commercial, and residential areas. In the total energy generation of the world, only 31.1% of energy is generated using renewable energy. The research in the field of EMS for a renewable energy system is on significant development. The need for reducing emission in generating energy makes a way for developing sustainable techniques and utilizes renewable energy sources.

1.7.5 RENEWABLE ENERGY SOURCES UTILIZATION IN EMS

Among the renewable energy sources, solar energy is the cleaner, inexhaustible energy resource. Solar energy is utilized in many ways including solar heater, solar PV, etc. Solar heater is used in domestic application because of easy installation. The solar PV and solar concentrators are used for power generation. The power generation requires large scale investment for bulk power needs. Solar energy is utilized in two ways solar thermal, converting sunlight into thermal energy and generating electricity and solar PV, directly generating electricity from sunlight. Solar energy is used in many places due to the abundant availability of sunlight and low maintenance. Solar energy is available only during the daytime and hence requires energy storage. The energy storage systems require charge controllers to protect them from overcharging and discharging. Wind energy is another renewable energy source utilized in large scale and medium scale applications. Electricity can be generated from a wind speed of 2–15 m/s.

1.8   FUTURE CHALLENGES AND SCOPE

In this section, key challenges and opportunities in facing CPS aspect of the SG are in view of ecosystems, big data, cloud computing, Internet of Things.

- Ecosystem View: SG development is always associated with the environment and social system. The nature, environment, and ecosystem are the flora and fauna, climatic changes, which are affected by improvement in the SG.
- Big Data: Big data is used in the data gathering and analytics. The five main aspects of big data are volume, velocity, veracity, variance, and value.
- Cloud Computing: In SG, the distributed resources in real-time management have to be met in a timely manner. Cloud computing is the paradigm with services such as computation, network, and storage act as resources. It has the advantages of self-services, pooling of resources, elasticity and increases security and solves privacy issues.
- Internet of Things: Internet of Things (IoT) is the extension of the Internet services due to the propagation of RFID, sensors, smart devices, and "things" on the Internet. IoT grows rapidly and is expected to be 50 billion devices connected to the internet by 2020. The development in IoT leads to the advancement in IoE.

1.9   CONCLUSION

The smart grid environment with EMS plays a significant role within the sensible utilization of electricity and demand response. The smart SGEMS with wireless networks and smart sensing element technologies elevates the standards of SG. Within the recent years, SGEMS has been considerably used and gains popularity due to high accessibility and convenience. The modern smart grid infrastructure with two-way communication, metering and observation devices paves the requirement for smart SGEMS applications. In future, the intensive use of SGEMS can amend the method of electricity usage and renewable energy utilization within the power network. Alternative energy may be the main contributor in renewable energy applications, while wind, biomass contributes comparatively less due to geographical and climate factors. The employment of renewable energy demonstrates the energy savings might be achieved from transmission energy losses and traditional installation. The design of CPS is challenging than coming up with of physical and cyber parts one by one. For CPSs, the required behavior of machine parts must be laid out in terms of their influence on the physical surroundings. Hence, a unifying framework is needed for modeling, which permits consistency and a low-overhead style.

KEYWORDS

- **power system**
- **cyber physical systems (CPS)**
- **cyber physical energy systems (CPES)**
- **renewable energy**
- **smart grid**
- **intelligent systems**
- **Internet of energy (IoE)**

REFERENCES

1. Amin, M. Minimizing Failure While Maintaining Efficiency of Complex Inter-active Networks and Systems: EPRI and US Department of Defense Complex Interactive Networks/Systems Initiative; First Annual Report, 2000.
2. Haase, P. Intelli Grid: A Smart Network of Power. *EPRI J.* **2005,** *27*, 17–25.
3. Profiling and Mapping of Intelligent Grid R & D Programs. *EPRI* **2006.**

4. European Smart-grids Technology Platform: Vision and Strategy for Europe's Electricity Networks of the Future. Directorate-General for Research Sustainable Energy Systems, 2006.

5. Davis, C. et al. Scada Cyber Security Testbed Development. In *NAPS*. IEEE, 2006; pp 483–488.

6. Schneider, K. et al. Assessment of Interactions Between Power and Telecommunications Infrastructures. *IEEE TPWRS* **2006**.

7. Sun, Y. et al. Verifying Noninterference in a Cyber-physical System the Advanced Electric Power Grid. In *QSIC*; IEEE 2007; pp 363–369.

8. Karnouskos, S. Cyber-physical Systems in the Smartgrid. In *INDIN*. IEEE, 2011; pp 20–23.

9. Giani, A. et al. The Viking Project: An Initiative on Resilient Control of Power Networks. In *ISRCS*. IEEE, 2009; pp 31–35.

10. Mo, Y. et al. Cyber–physical Security of a Smart Grid Infrastructure. *Proc. IEEE* **2012**, *100* (1), 195–209.

11. Susuki, Y. et al. A Hybrid System Approach to the Analysis and Design of Power Grid Dynamic Performance. *Proc. IEEE* 2012.

12. Saber, A.; Venayagamoorthy, G. Efficient Utilization of Renewable Energy Sources by Gridable Vehicles in Cyber-physical Energy Systems. *Syst. J. IEEE* **2010**, *4* (3), 285–294.

13. Zhu, Q. et al. Robust and Resilient Control Design for Cyber-physical Systems with an Application to Power Systems. In *CDC-ECC*, 2011.

14. Hadjsaid, N. et al. Modeling Cyber and Physical Interdependencies-application in ICT and Power Grids. In *IEEE/PES PSCE*, 2009; pp 1–6.

15. Zhao, J. et al. Cyber Physical Power Systems: Architecture, Implementation Techniques and Challenges. *Dianli Xitong Zidonghua (Autom. Electric Power Syst.)* **2010**, *34* (16), 1–7.

16. NIST Special Publication 1108R2. NIST Framework and Roadmap for Smart Grid Interoperability Standards, Release 2.0, 2012; http://www.nist.gov/smartgrid/upload/NIST Framework Release 2-0 corr.pdf

17. Liu, Y.; Ning, P.; Reiter, M. False Data Injection Attacks against State Estimation in Electric Power Grids. *ACM TISSEC* **2011**.

CHAPTER 2

An Intelligent Traffic Management System

M. GOWTHAM[1*], M. K. BANGA[2], MALLANAGOUDA PATIL[2], and
NATARAJAN MEGHANATHAN[3]

[1]Department of Computer Science and Engineering, National Institute
of Engineering and Institute of Technology, Mysuru, Karnataka, India

[2]Department of Computer Science and Engineering,
Dayananda Sagar University, Bangalore, Karnataka, India

[3]Department of Computer Science, Jackson State University, United States

*Corresponding author. E-mail: gouthamgouda@gmail.com

ABSTRACT

Traffic gridlock, congestion, or commonly Traffic jam is likely to be a
serious problem in most of the countries. There might be several reasons
for this. If we need to list a few failures in the traffic signals, the poor
enforcement of the law, and the old unmanaged traffic management system
and lack of availability of resources for the expansion of the infrastructure
are some reasons for the traffic congestion. In today's metropolitan life,
people spend most of the time in the husky, congested traffic and the toll
plazas wasting their valuable time. By this, people not only waste their
time, but also are affected by the pollution, and they are prone to many
health disorders. There are quite a few previously defined methods from
which the solutions can be found, but they are not up to the expectations of this fast-growing world. The main objective of this chapter is to
enable a system that runs on the principle of Artificial Intelligence and the
Internet of Things to manage the traffic signals in metropolitan and major
cities. This system is capable of determining the overall vehicle density

on road at run time in every direction and dynamically allocates the signal timings at the junctions. Another system that works and the principle of sound recognition to manage the traffic on the siren sound of emergency vehicles. Finally, a fully automated system using the GPS service to get the current traffic status and decide the most suitable path intelligently. The above methods are integrated together to form a high-end traffic management system.

2.1 INTRODUCTION

In today's expanding and fast-growing world, human has completely relied on his life on the machines mostly for communication and transportation. This growing need has made humans invest on machines that may help them work more smoothly and precisely than before. These growing-needs have also bought a lot of problems for them knowingly or unknowingly. Traffic congestion is one of those. Traffic congestion occurs mainly due to slower speeds, increased queuing of the vehicles, increased trip time, and also when the number of vehicles on service exceeds the carrying limit of the road. These cases may influence to dreadful conflicts such as collisions or abrupt braking of a vehicle which may additionally induce the steady flow of traffic to have rippling outcomes. There do exist other collateral quandaries in traffic systems due to anti-social factors, which direct to stagnation of traffic at a single place. An average of Rs. 150,000 crores goes waste due to congestion. The reasons are the reduced propagation speeds of freightage vehicles and prolonged standing time at checkpoints and customs plazas. Considering the main corridors of the countries, the average speed of the freight vehicles has reduced at least by 25%. Considering the rate of growth of traffic, the average speed has reduced to 20 to a maximum of 35 km/h.

According to[1] 49% of residents in a metropolitan or other major city spend almost 12 h (half a day) or over 100 min every day in the traffic. We can conclude that the average human spends a quarter of his lifetime in the traffic. This time can be utilized in the overall development of a person or the country. So, there is a high demand for a traffic management system that can be relied on and on the other hand must provide better performance to efficiently manage the traffic. In the fast-growing world of today, people are expanding their facilities without any consideration toward the environment leading to a lot of destruction to the environment and lack of technology to at least reduce the damage done to the surroundings and that's where traffic management steps in if not all the damage at least a portion of the damage done to the surroundings can be reduced using traffic management signals and technologies. This high volume of traffic in cities has brought in many problems. Just to list a few of them like an increase in the number of accidents that are fatal and also health issues like skin cancer, dust allergy, and breathing problems which may be dangerous in certain situations. This has also blocked the ways for emergency vehicles like ambulance, police or the fire-extinguishers when on duty. This may lead to a risk in the life of the human. In today's hectic life, the people of our society have rarely given any thought to the surroundings and for that same reason we often neglect the effects that are caused by our reckless actions to our society. Considering all the facts, their pros and cons, we need a certain system or methodology designed to cope up with all these non-human friendly hacks. So, there is a great demand to regulate traffic in an intelligent way as the regulation of traffic with the customary or the traditional way such as the signaling system or other common techniques are not having a significant effect on curbing congestion of vehicular traffic.

2.2   RELATED WORK

The traditional or the conventional methods of traffic management were not adequate in order to adapt to the expanding vastness of the technology and its slowly growing power, therefore, the new aspects required a giant leap in the development of the traffic management systems.

2.2.1   CONVENTIONAL METHOD

The old and traditional method of traffic management is signaling using tricolors. Here, the traffic signal timings are set at the time of installation of the device. This has many disadvantages that are harder to work upon and these devices are not so reliable and lack performance. In this, the timings remain the same almost throughout the lifetime of the device and if in case it should be altered, a long process has to be followed which is a time-consuming process. The main drawback of this system is the static preset timings for the traffic signal which are changed hardly in a year. As there is a hard increase in traffic, the demand for traffic management has also increased. New methods are needed for the purpose of this increasing traffic demands. This type of system also requires a high maintenance cost. The traditional or the conventional methods of traffic management were not adequate in order to adapt to the expanding vastness of the technology and its slowly growing power, therefore, the new aspects required a giant leap in the development of the traffic management systems, therefore, further new branches were introduced in order to attain proper advancement in the technology related to traffic management. Although the traditional technology was good, with the giant leaps and bounds of technology. So it is necessary to adopt new technologies. New technologies that are more advanced and reliable that can work efficiently and should have fewer disadvantages.

2.2.2   *VISUAL LIGHT COMMUNICATION (VLC) METHOD*

The need for a good traffic management system has given rise to many new technologies. One of these is the visual light communication (VLC) based traffic management. According to,[2] this system uses the electromagnetic frequency produced in the visible light is used to manage traffic. In this system, the vehicle headlamps can communicate with the traffic signals and this can be helpful in managing the traffic, but this technology is still in its development stage. The main drawbacks of this type of system are the intensity level of the vehicle headlamps and also the obstacles that may come in the path of the optical signal may hamper the proper functioning of the system. The system would fail to work at daytime as turning on the headlights at the day time may lead to energy loss. The VLC method of traffic management was a huge development in the traffic management corridor but it is still a work in progress, therefore, for implementation a 100% assurance is required on the project.

2.2.3   *INDUCTIVE LOOP DETECTION METHOD*

Inductive loop detection operates on the principle that larger turns of insulated wire are placed in an inconsiderable depression in the roadway, a lead in wire runs from the roadside pull box to the controller and to the electronic assembly located in the controller cabinet. When a vehicle crosses over the loop or stops, the induction of the wire is modified. Due to the difference in induction, there is a variation in the frequency. This difference in the frequency prompts the electronic unit to transfer a signal to the controller; symbolizing the presence of the vehicle.[6] Inductive loop detection is helpful in knowing the vehicle proximity, passage, occupancy, and even the total number of vehicles passing through a precise area.[7] But, there are a few drawbacks with this system. These include reduced reliability due to improper connections made in the pull boxes and due to the application of sealant over the cutout of the road. If this system is implemented in the poor pavement or where renovating of the roads is frequent then the problem of reliability is intensified.[6]

2.2.4   VIDEO ANALYSIS METHOD

The video analysis system consists of an intelligent camera connected which consists of sensors, a processing assembly and a communication system.[3] The traffic is continuously observed using a smart camera. The video interpretation abridgments view description from the rough video data. This description is then used to estimate traffic statistics. This statistic involves the frequency of the vehicles, the average speed of the vehicles as well as the lane usage.[9,10] The drawbacks associated with video analysis systems are the overall cost of the system is pretty high, the system gets affected in case of dense fog or heavy rains and night-time surveillance needs proper lighting.[3]

2.2.5   INFRARED SENSORS METHOD

The energy transmitted from road surfaces, vehicles, and other objects is detected using the infrared sensors. This technique can be practically implemented by adopting an optical system that can transform the energy obtained by these infrared sensors which are in turn directed onto an infrared sensitive material, finally into the electric signals. These infrared sensors can be used for the detection of foot-traveler in crossings, the transmission of traffic data and signal control.[4] These sensors are positioned aloft to inspect the traffic and most commonly attached to the streetlights and the signal towers. The central drawbacks of using the infrared sensors are that the establishment and maintenance of the system are tiresome and the performance of the system may get altered due to fog, haze, or showers.[4]

2.2.6   *USING RADIO FREQUENCY IDENTIFICATION*

In this, each vehicle is fitted with an radio frequency identification (RFID) tag. This RFID tag here stocks all the data regarding the vehicle such as the vehicle owner's name, vehicle number. These tags can be used for recognizing each vehicle uniquely and also assist the driver to obtain certain traffic information directly from the server. Here, the current traffic signaling system is linked with the RFID controller which has the data concerning with every vehicle that crosses by it. Thus, when a vehicle crosses by a signal, the controller automatically keeps the record of the vehicles passing by it. Using these records, the traffic signals are manipulated. The processes of switching the signal ON or OFF depend upon the signal data sent by the RFID controller. The main drawback is the cost of installing RFID tags to every vehicle and it is practically impossible for the existing vehicles, as there are a huge number of vehicles present today.

2.3   HYBRID TRAFFIC MANAGEMENT SYSTEM

Integration of microwave radar with the AI system which is in turn coupled with the embedded IoT systems.

2.3.1   *SYSTEM OVERVIEW*

The system is simple, small, and elegant in design and can work more precisely and more efficiently. The system mainly consists of a central database, along with the radar system which is coupled with the computing unit and the transmitting unit. These systems are then integrated and they are placed at an average distance of 300–500 m between each and also from the actual signal junction. The microwave radar is used to detect the number of vehicles crossing the radar path in a particular amount of time and in a particular unit area. The data collected by the sensors are transferred to the computing unit which is a system built under artificial iIntelligence using several algorithms. The data are processed by taking into account of the total number of vehicles passed through the benchmark sensor; the density of vehicles is calculated and is categorized as low, medium, and high. If the number of vehicles passing is 5–10, the density is set to low and in the case of medium it ranges from 15 to 25, where 30 and above are considered to be high density. There may be variations in the range but the AI system is compatible to analyze the total number and make decisions accordingly. These readings are synchronized to the database continuously from where the data are retransmitted to the signal junction location. Based on the value obtained from the database, the AI system which is connected with the hardware signaling device performs the operations and sets the signal timings and also it manipulates the signaling pattern. These timings and pattern change at real time based on the values from the database which may be low, medium, or high. If the data received is low, the signal timings may range from 20 to 25 s, and the traffic moves at a considerably faster rate. The values for medium ranges from 30 to 45 s and for that of high, it ranges from 60 to 90 s. These radars have high working efficiency and are not distracted by the fog and heavy rains. These have high range of working frequency and hence can work more precisely (Fig. 2.1).

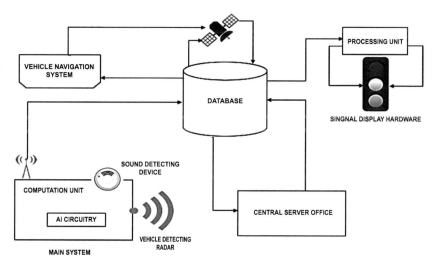

FIGURE 2.1   System prototype.

2.3.2 WORKFLOW EXTENSION

The above-mentioned system can be enhanced to get more benefits from the same devices. Generally, the sirens of the emergency task force are designed in such a way that they produce and project high pitch sounds toward the forward direction which is an application of Doppler's effect. They are designed in such a way that the sound seems to be nearer while the actual source originating the sound is quite far. Using this effect and phenomena, the traffics can be cleared if in case any of these vehicles is emerging. The ideology used here is quite simple. The sound sensors which are designated to detect siren sounds are also coupled with the main system which on receiving input signals generates different dataset and transfers it to the computation unit and the data are transferred to the database and n turn to the signal location where the signal timings are altered according to the direction of the emergency vehicle.

Further extensions can be achieved by making the use of the data in the database more efficiently. Most vehicles today come inbuilt with a navigation system. These navigation systems can communicate with the main database by establishing a secure connection and follow the routes which are not so dense. The database has real-time data entry of each signals and systems installed. The emergency vehicles can also make use of these services so that they can choose a shortest path and also the less dense road. The system may also be used as an advanced level green corridor manager. Green Corridors are basically zero traffic roads which are implemented at times when VIP's travel. This system comes in handy as the implementation is easier and the time consumed will be relatively less (Fig. 2.2).

2.4 APPLICATIONS

The proposed system has some of the applications.

2.4.1 AUTOMATIC SPEED DETECTION

No extra systems are needed as this system is capable of performing multiple tasks. This system can be used to determine the speed of a motorist and to detect if they violate the designated/set speed limit. If the motorist outrages the rule, the penalty will be calculated in the server and billed monthly to the vehicle owner.[5]

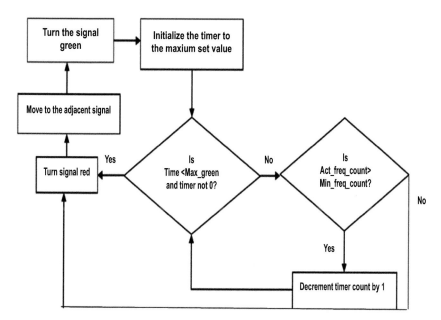

FIGURE 2.2   System working algorithm.

2.4.2   REDUCED RISK

This system can help in the reduction of the risks that are prone to the persons who are in the traffic in both relative and absolute ways. The system is persistent to implement measures through which the pedestrians are at a low risk as walking on the footpaths is also dangerous in these days. Apart from the trespassers, the people who board the vehicles also need to be taken care of. The vehicles when managed, manipulated to move in certain paths with calculated movement in several directions can lead to risk reduction.

2.4.3   CONTROLLED FLOW OF TRAFFIC

This system ensures that there is no blockage for the emergency service vehicles and also ensures the free and easy flow of the traffic without any lags in between. This almost reduces the traffic congestion problem to the level zero virtually. Further improvements in the law and the system may bring out more effective results.

2.4.4   ROAD USAGE STATISTICS

This system also keeps track of the number of vehicles using that particular road and develops statistics on the basis of usage. This statistic can be used to determine the type of pavement required based on the number of vehicles and also the type of vehicles that are using the current service.

2.4.5   DETECTION OF TRAFFIC VIOLATORS

This system can also recognize some of the traffic violations such as the vehicles on the wrong way and also detect vehicles in the no-entry zone.

2.5   CONCLUSION

The proposed work focuses on the Hybrid Traffic Management. This chapter presents the real-time vehicle detection and estimating the vehicle density on a road which will reduce the shortcomings of the present working system such as dependency on the environmental conditions, high implementation cost. The designed system aims at developing cost–effective methods than the existing system and also effective management of traffic congestion. This is done using key technologies such as the internet of things and artificial intelligence based on the sensor reports. Also, this detailed report can be used to create perfect or adequate plans for vehicle detection and vehicle density by taking into account the various factors that affect the environment and with suitable precautions kept in mind as well in recent days, the traffic all around the world has skyrocketed which has resulted in a severe drop of the thickness of the ozone layer leading to global warming at alarming rate examples of this are the frequent forest fires due to high temperatures. By using AI, the developments are quite surprising not only do we use AI in robots for various purposes, but we also use AI for proper traffic management as well. The different traffic management systems gave us an insight into how the day-to-day development of the traffic management is taking us, and it shows that all the eats that were considered as an impossible task in the past are becoming a reality in today's life as due to the drastic development in the technology. Although it is impossible for us to say that we are at the pinnacle of our development in traffic management. We can say that we are nearing the destination in traffic and it is not too long before we achieve our goal of reaching the point where the complete development of the vehicle density-based traffic management. We are also using various other branches to assist us in the quest for excellence in this field like corridor projects, VLC, and etc.

Besides, this chapter also manifests the challenges faced in metropolitan zones across the world induced by congestions and relevant sources. Overcrowding originated a problem, which has a great impact on economies worldwide. Particularly, cosmopolitan zones are worst hit following these circumstances. Blockages have unfavorable repercussions on the economic and financial standings of a country, also on the surrounding environment and consequently, reduce the overall quality of living. The introduced method can be intensified by applying some other dominant communications or information networks mediums other than microwave radars.

KEYWORDS

- **traffic management**
- **internet of things**
- **artificial intelligence**
- **sensors**
- **metropolitan cities**

REFERENCES

1. Dash, D. K. India loses Rs 60,000 Crore Due to Traffic Congestion: Times of India, TNN May 31, 2017. http://articles.timesofindia.indiatimes.com/2012-05-31/india/31920307_1_toll-plazas-road-space-stoppage

2. Kumar, N. Visible Light Communication Based Traffic Information. *Int. J. Future Comput. Commun.* February 2014.

3. Wu, B.-F. A New Approach to Video-based Traffic Surveillance. *IEEE Intelligent Traffic Society*, March 2013. http://ieeexplore.ieee.org/xpl/articleDetails.jsp?arnumber=6264098

4. Hussain, T. M. Overhead Infrared Sensor for Monitoring Vehicular Traffic: Vehicular Technology. *IEEE Transactions* **1993,** *42* (4), 0018–9545.

5. Xu, Y.; Jin, Y. Remote Road Traffic Data Collection and Intelligent Vehicle Highway System. Jun 4, 2002, US Patent 6,401,027.

6. Radio Frequency Identification (RFID) Controller. US7245220 B2. Jul 17, 2007. http://www.google.com/patents/US7245220

7. George, B.; Vanajakshi, L. A Simple Multiple Loop Sensor Configuration for Vehicle Detection in an Undisciplined Traffic Sensing Technology (ICST). *Fifth Int. Conf.* **2011,** 21568065. http://ieeexplore.ieee.org/xpl/articleDetails.jsp?tp=&arnumber=6137062&url=http%3A%2F%2Fieeexplore.ieee.org%2Fxpls%2Fabs_all.jsp%3Farnumber%3D6137062

CHAPTER 3

Real-Time Monitoring and Tracking of Intrusions in Warehouses Using Video Analytics Techniques

JEEVA S.[1*], SIVABALAKRISHNAN M.[2], SARANGAM KODATI[3], and HONG-SENG GAN[4]

[1]*Department of Data Science and Business Systems, SRM Institute of Science and Technology, Kattankulathur, Chennai, Tamil Nadu, India*

[2]*Vellore Institute of Technology-Chennai Campus, Chennai, India*

[3]*Brilliant Institute of Engineering and Technology, Telangana, India*

[4]*Universiti Kuala Lumpur, British Malaysian Institute, Gombak, Malaysia*

Corresponding author. E-mail: sassyjeeva@gmail.com

ABSTRACT

Automated intrusion detection in the warehouse is a more challenging task in real-time era. The algorithms available in literature are suitable for stored videos but not for real-time videos. Very few algorithms are work in real-time but they took more computation power so that is not adapted by cyber-physics hardware. In this chapter, some video analytics techniques are discussed, compared the results, and explained suitable selection of algorithm for an application. In video surveillance, all algorithms are not suite for all application. Most of the algorithms are environment specific which means algorithm design for a specific environment and it acts best in that situation. Some algorithms are work better on most of the environment in the change detection dataset 2014. The dataset consists of 12 categories of videos and each category has more videos for evaluation of the proposed algorithm. The universal algorithm for motion detection is a visual background extractor. It gives the best result in most of the videos in the change detection dataset 2014. And also some cyber-physical hardware with their use and how to interface with algorithms has been discussed in this chapter.

3.1 INTRODUCTION

Stockrooms and dissemination focuses are regular focuses of robbery, burglary, and pilferage. Security dangers originate from both inward and outer gatherings. Interior dangers are presented by representatives and outsiders enlisted by the association, while outer dangers would include any other individual who enters the distribution center without approval. There are different ways the security threats may occur by associated people, and facility limits: Use separate zones for accommodating merchandise[4] and shipping. But there is an illusion on the broader of merchandise and shipping. Introduce an obstruction, for example, a fence around the outside yard of the distribution center. Keep the door secured constantly at the time of the distribution center is closed. In the event that fundamental, chance directors may consider keeping the entryway bolted consistently and just enabling access to approved representatives. All the entryways must have traceable methods and all entryways when the distribution center is shut.

Visitor sign-in registers: Establish a guest register so everybody who enters the stockroom is recognized. The employees are not allowed the outside intruder to associate with them to enter the stockroom with either guest or drivers.

Reconnaissance outlines and electronic security: the current available electronic device has only the access limit up to either room or their limits. The framework[5] available for entrance monitoring and controller can collect the entire information about the intruder such as when up to what time, and what esteem. The video surveillance framework has covered a wide area and has more information about the area of interest. The focus cameras have set a view of view passageway for looking just as classified territories.

Employee checks: The above techniques will ensure a distribution center against outside dangers; however, not those introduced by inner gatherings. Representative burglary can make stunning misfortunes inside distribution centers. Direct careful personal investigations of distribution center workers before employing, giving specific consideration to any records of burglary or unexplained occupation misfortune in stockroom or capacity occupations. Set up a mysterious detailing framework. This will enable workers to report a colleague they accept is taking without dread of repercussion.

Among this means for verifying distribution center, reconnaissance[8] assume a significant job in security and gives more data contrast with different techniques. So the observation and digital-physical framework are inserted to make completely mechanized reconnaissance framework. The framework will work without human interlude and act at time crisis all alone.

3.1.1 CYBER PHYSICAL SYSTEM

The technology cyber-physical system[7,12] has a combination of both cyber and physical things. The people working in this technology have to clear understanding of cyber word and understanding about physical things. In CPS,[13] generally embedded computers then network monitor and then control of physical processes. The entire system acts as a feedback loop so any intercept in physical process can also affect the computation of the system and vice versa. CPS includes all these three parts. It also acts as both static and dynamic properties of all things.

3.1.2 VIDEO ANALYTICS

Video analytics[3] is defined as the analysis of video to get useful information from the videos.

It used to develop a lot of applications which can useful for the society. An application like surveillance, intrusion detection system, Monitoring, Internet of Things (IoT), home automation, automated car, video surveillance, traffic surveillance, and so on problem are come under this collection of expectations. In video analytics, the moving object detection field is still a dynamic area of research, and the best possible solutions have so far not been found. Although simple scenarios are reliably and exactly handled by existing methods, more difficult scenarios such as complex background, low-frame rate, dynamic background, and continue to be a challenge. The inspiration of this research is to get better strength of object detection, by integrating background model, environment-independent and limited storage in detection. The techniques proposed up to date for evaluation of object detection are lack in background model which means entire background information's are unable to fit in background model. It can be used for small video processing applications like weather forecast and financial market they are used a homogeneous type of variables so it processes the metadata and slight video processing techniques on the video. In previous video, storage is a more challenging task. So they compressed the video for object tracking and their aim to produce efficient and robust tracking of objects. They extend the study to behavior analysis, unusual behavior, IoT, emotional analysis, crowed analysis, and so on. Visual object tracking covers most of the techniques to aim the efficiency but still required the improvement and optimal solution to detect and track the object.

3.1.3   IOT

IoT represents Internet of Things. The principle elements of IoT stages are fivefold: associate, secure, oversee, dissect, and manufacture.[14] Each fold can be useful for adapting IoT: associate data frameworks, information has secured inside stage and the interfaces,[15] achieve the outlines and procedures associated, investigate the information that is inside the stage to give important data, and fabricate applications and specialists to help client choices.

IoT stage shows propelled aptitudes on security and investigation fields as shown in Figure 3.1. An investigation is purely depends on each stage namely, prescient, subjective, continuous, and con-literary techniques. Skeleton and distributed evaluation highlights on additionally included in the investigation tool[14] at the same time optical examination and machine learning are likewise maintained.[17] while the territory of security, the stage utilizes TLS convention and furthermore conveys examination capacities for the general wellbeing of the framework, for example, recognition of patterns and distinguishing proof of abnormalities.[16] The security investigation capacity conducts examination only for higher security of the general framework.

FIGURE 3.1   General architecture of IoT.

3.2   LITERATURE REVIEW

In overview, there are various proposed foundation demonstrating techniques, which are specific to essential models, factual models, neural system model, group model, estimation model, and progressed measurable models. This is generally in light of the way that no single technique can adjust to all of the challenges around there. There are a couple of issues that a fair foundation subtraction count must assurance. At first, it must be solid against changes in light. Second, it should keep away from recognizing non-stationary foundation objects, for instance, affecting leaves, grass, downpour, snow, and shadows thrown by moving items. Finally, the foundation model should be delivered with the ultimate objective that it needs to react quickly to changes far out, for instance, the start and end of vehicles.

3.2.1   VIDEO ANALYTICS ALGORITHMS

Video analytics are self-governing comprehension of occasions happening in a scene observed by different camcorders, has been quickly advancing over the most recent two decades.[1] Video investigation likewise makes reconnaissance frameworks increasingly smart to decrease huge measures of picture information to sensible dimensions. Canny video observation frameworks can for instance naturally dissect and label reconnaissance video continuously, recognize suspicious exercises, actuate cautions or different activities to alarm administrators or other faculty and start video recording.

The time differencing method[6] is the basic system for transformation or movement recognition. Object transformation between the pictures can leads to modification on the edges and updated by model straight-forward contrast on pixel-wise premise and looking at the incentive from a pre-characterized edge. For useful frameworks increasingly thorough foundation and closer view of other techniques are required, which are all outlined in Figure 3.2 for various strategies. Thus, the foundation turns out to be progressively unique the multifaceted nature of method develops on basic Gaussian Mixture Model (GMM) to mixture of Gaussian (MOG) to surfaces and digital stream strategies.

3.2.2   REAL TIME ALGORITHMS

The methods available for real-time video processing[1] for identification of foreground detection are very limited and very few algorithms can support the entire environment. The algorithms are Visual Background Extractor Method (ViBE), Pixel-Based Adaptive Segmenter (PBAS), and Self-Organizing Background Subtraction (SOBS). These methods are satisfying most of the environment and developing the background model for foreground detection.

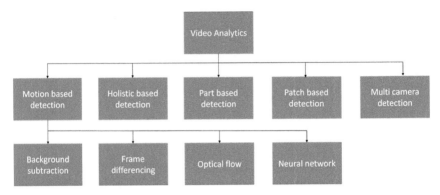

FIGURE 3.2   Methods in video analytics.

The past strategies still have numerous unsolved issues in building a productive picture and video object identification and recognition framework. These issues prompt an unsatisfied execution of both object discovery and recognition in real-time applications. There are some problems identified in building an efficient and robust extraction/subtraction system:

Presently, available algorithms for extracting the background from images deal only with non-moving pixels. No algorithm has so far detected the background form of various kinds of images to deal with a moving pixel. In many algorithms for background subtraction, the reference image is only a prior assumption. In all the methods the shadow effect is not discussed.

Hence, an effective unified and robust Background Subtraction system is required to distinguish the background image and foreground image, which take care of the moving pixel, shadow effect, and illumination, so as to make the system suitable for heterogeneous images.

3.2.3   ROLE OF CPS IN VIDEO ANALYTICS

Sensors[11] get input from the scene. By adding the hub investigation to the blend, information examination can be improved by dispensing with correspondence with the cloud. Distributed computing requires two if not three requests of an extent of more data transmission than hub examination applications. Therefore, hub examination expends significantly less computational power and dormancy is diminished. Thickly populated markets, turbulent traffic segments, and city parking spots are a portion of the mind-boggling airs that can be detected for prescient and social investigation utilizing hub examination. Abnormal state handling of these conditions performed in the cloud can propel business techniques, redirect general traffic stream, and improve the viability of government-oversaw parking spots. Notwithstanding, by actualizing lower-level programming at the sensor hub as opposed to performing investigation in the cloud – inertness, data transfer capacity, security, and power can be improved in these situations. The resultant output pass to the relay and then action to be performed by the actuators as shown in Figure 3.3.

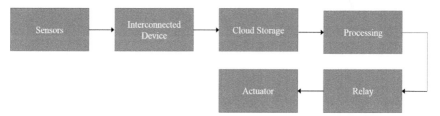

FIGURE 3.3 Architecture of Cyber-physical system.

3.3 ARCHITECTURE OF INTRUSION DETECTION

The overall architecture of the intrusion detection has been given in Figure 3.4.

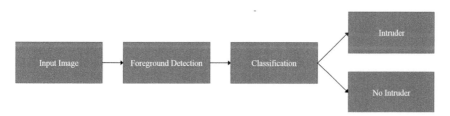

FIGURE 3.4 Overview of intrusion detection system.

Input video[10] has been processed to identify the foreground from the current image. The detected foreground image has passed to classifier for classification of intruder or not.

3.3.1   PHYSICAL SYSTEM

The physical system consists of hardware and sensors to get input and preform output on real-time environment. It acts as an input by sensing the area and send to the cloud. After processing, the output has sent it to the actuator to perform corresponding operation to be carried out. Physical and logical states are tightly coupled and each one interdependent.

3.3.2   LOGICAL SYSTEM

A logical system means the processing of physical data.[9] The data which are collected from the physical device are processed in logical system such as in cloud, IoT device, and local computers.

The sensor gets the data or video information that can be stored and processed by the logical state and return the output to the actuator to perform the operation. Both states are interconnected and feedback system. So any interruption occurred at any state will reflect to another state also.

3.4   VIDEO ANALYTICS IN IOT

Video examination applications working inside the IoT is shown in Figure 3.5. It can be improved by utilizing hub investigation and logarithmic imagers. Video examination[2] applications endeavor to exploit the inexhaustible data in the regular world for a few reasons. These reasons go from facial acknowledgment to ordinary reconnaissance; however, the greater part is fixated on prescient and conduct investigation. The data accumulated in these applications can be prepared widely at a more elevated amount by distributed computing. Be that as it may, top-to-bottom handling has its confinements and can be improved from multiple points of view by including hub investigation and logarithmic imagers to the blend.

3.4.1   GENERAL ALGORITHMS

The common issues in background subtraction approaches are noise, moving background, and illumination changes. So the background has to change over a particular period of time in an arrangement. Therefore, scientists started to redevelop the background subtraction method by introducing adaptive background models. It updates the background model more frequently at a particular time interval. So system updates the illumination changes on the background model.

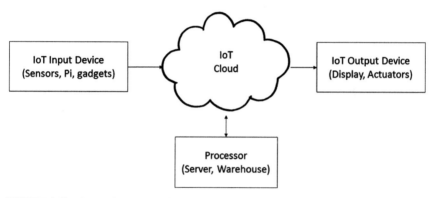

FIGURE 3.5 General structure of IoT.

$$D(x, y) = \begin{cases} 1 & |I_k(x,y) - B_k(x,y)| > \tau \\ 0 & \text{Otherwise} \end{cases} \tag{3.1}$$

where $D(x,y)$ is the binary difference image between the background image $B_k(x,y)$ and the current image $I_k(x,y)$ at the kth frame with respect to time in video sequence.

If the camera has in motion, then the background has updated in dynamically. So the method has to update more frequently for identifying the foreground from the current image.

$$B_{(k+1)}(x, y) = \alpha I_k(x, y) + (1 - \alpha) B_k(x, y) \tag{3.2}$$

where α is an adaption coefficient chosen arbitrarily. The α is greater than it faster changes in background image but still it has lot of struggles in the detection foreground object and sudden illumination changes.

Apart from these basic algorithms, some standard algorithms are also there for the detection foreground image from the frame. The algorithms such as Principle Component Analysis (PCA), Support Vector Machine (SVM), Robust Principle Component Analysis (RPCA), GMM, SAMple CONsense (SAMCON), Deep Neural Network (DNN).

3.4.2   REAL-TIME ALGORITHMS

The foreground detection methods are available for real time such as ViBE, Mixture of Gaussian model, PBAS, SOBS, and Twin background subtraction. These methods have lot of research gap to enhance the method for further improvement on performance of the algorithm. The past strategies still have numerous unsolved issues in building a productive picture and video object identification and recognition framework. These issues prompt an unsatisfied execution of both object discovery and recognition in real-time applications. There are some problems identified in building an efficient and robust extraction/subtraction system:

- Presently, available algorithms for extracting the background from images deal only with non-moving pixels. No algorithm has so far detected the background form of various kinds of image to deal with a moving pixel.
- In many algorithms for background subtraction, the reference image is only a prior assumption. In all the methods the shadow effect is not discussed.
- Hence, an effective unified and robust Background Subtraction system is required to distinguish the background image and foreground image, which take care of the moving pixel, shadow effect, and illumination, so as to make the system suitable for heterogeneous images shown in Figures 3.6 and 3.7.

FIGURE 3.6   Input and output of SOBS method.

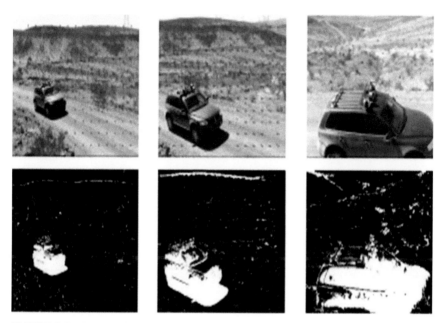

FIGURE 3.7 Input and output of visual background extractor method.

3.4.3 ALGORITHM SUIT IN IOT

All algorithms are not suit for IoT application because they required more computation power and storage capacity. But embedded device has less computation power and static memory which is not expandable. So algorithm required to optimize for the compatible of IoT device.

The algorithms are ViBE, TBGM, and MOG to have a static memory and required very less computation power.

3.5 COMPARING RESULTS OF EXISTING ALGORITHMS

The algorithms has compared with different video dataset for the evaluation. The results are gathered from the different method and found that TBGM has better performance in dynamic background dataset shown in Figure 3.8. The reason behind is continuously update the pixel value to the background model. The method performed well on PTZ dataset also because it has a feedback system on algorithm so every pixel has feed in background model.

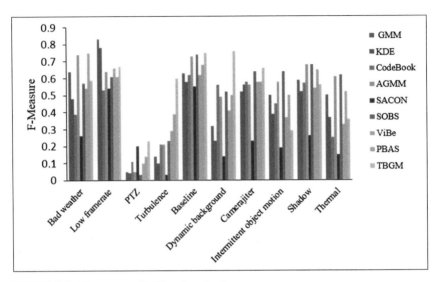

FIGURE 3.8   F-measure of different methods.

Further turbulence, baseline, and camera jitter dataset also produce considerable improvement in the result. But other dataset gave better result but still required some improvement in method may be done by increasing number of frames in long term background model. F-measure value can be compared with existing methods because most of the papers use this metrics for their evaluations. So, we also use it and produce the result to compare the F-measure value. From F-measure value, it shows that Twin background model have more accurate on most of the environment in change detection dataset. Figures 3.9 and 3.10 show the output of different algorithms and test on different videos in change detection dataset.

3.6   CONCLUSION

Currently, lot of research has been done for moving object detection using video analytics techniques. At certain situation, it is very hard to identify the moving object and it is very difficult to execute the algorithm on the embedded device at real-time application because it has very limited computation power and memory. The method also supports batch processing in moving object detection algorithm. Despite of people, the

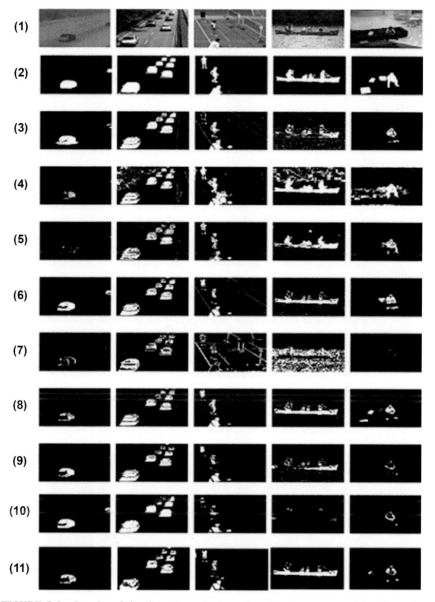

FIGURE 3.9 Results of the datasets on bad weather, baseline, camera jitter, dynamic background, intermittent object motion. (1) Original frame, (2) Ground truth, (3) GMM, (4) KDE, (5) codebook, (6) AGMM, (7) SACON, (8) SOBS, (9) VIBE, (10) PBAS, and (11) TBGM.

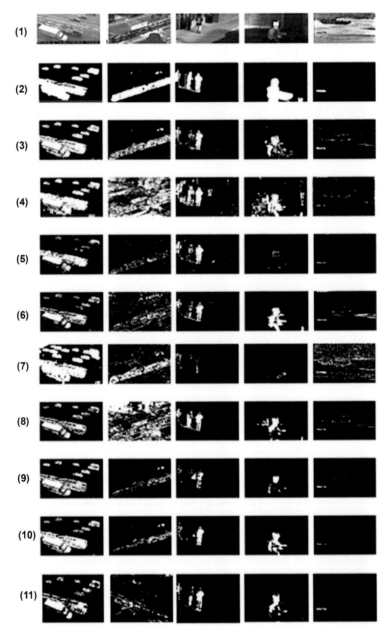

FIGURE 3.10   Results of the datasets low frame rate, PTZ, shadow, thermal, turbulence. (1) Original frame, (2) Ground truth, (3) GMM, (4) KDE, (5) codebook, (6) AGMM, (7) SACON, (8) SOBS, (9) VIBE, (10) PBAS, and (11) TBGM.

method identified as toy, stickers, etc. has a human while people entry and exit from the shop. The algorithm available in literature is not only used to CPS for the warehouse, but it also gives more accurate results at the same time, false alarm also high and also optimization of existing algorithms gives better result on cyber-physical system in the warehouse intrusion detection.

KEYWORDS

- **video analytics**
- **computer vision**
- **IoT**
- **cyber physical system**

REFERENCES

1. Jeeva, S.; Sivabalakrishnan, M. Survey on Background Modeling and Foreground Detection for Real Time Video Surveillance. *Elsevier Procedia Comput. Sci.* **2015,** *50,* 566–571.
2. Jeeva, S.; Sivabalakrishnan, M. Robust Background Subtraction for Real Time Video Processing. *Int. J. Pure Appl. Math.* **2016,** *109* (5), 117–124.
3. Jeeva, S.; Sivabalakrishnan, M. *Twin Background Model for Foreground Detection in Video Sequence.* Cluster Computing: Springer, 2017; https://doi.org/10.1007/s10586-017-1446-7
4. Lee, E. A.; Matic, S.; Seshia, S. A.; Zou, J. The Case for Timing-centric Distributed Software. In *IEEE International Conference on Distributed Computing Systems Workshops: Workshop on Cyber-Physical Systems*; IEEE, June 2009 [Online]; http://chess.eecs.berkeley.edu/pubs/607.html
5. Eidson, J.; Lee, E. A.; Matic, S.; Seshia, S. A.; Zou, J. A Time-centric Model for Cyber-physical Applications. In *Proceedings of 3rd International Workshop on Model Based Architecting and Construction of Embedded System (ACESMB 2010),* October 2010; pp. 21–35. [Online]; http://chess.eecs.berkeley.edu/pubs/791.html
6. Broman, D.; Derler, P.; Eidson, J. Temporal Issues in Cyberphysical Systems. *J. Indian Instit. Sci.* **2013** [Online]; http://chess.eecs.berkeley.edu/pubs/997.html
7. Cardenas, A.; Amin, S.; Sinopoli, B.; Giani, A.; Perrig, A.; Sastry, S. S. Challenges for Securing Cyber Physical Systems In *Workshop on Future Directions in Cyber-physical Systems Security. DHS,* July 2009 [Online]; http://chess.eecs.berkeley.edu/pubs/601.html

8. Barr, M.; Massa, A. *Programming Embedded Systems: With C and GNU Development Tools*, 2nd ed.; O'Reilly Media, 2006 [Online]; http://shop.oreilly.com/product/9780596009830.do

9. Lee, E. A.; Seshia, S. A. *Introduction to Embedded Systems—A Cyber-Physical Systems Approach*, 1st ed.; Lee and Seshia, 2010 [Online]; http://chess.eecs.berkeley.edu/pubs/794.html

10. Derler, P.; Lee, E. A.; Sangiovanni-Vincentelli, A. Modeling Cyber-Physical Systems. *Proc. IEEE* (special issue on CPS) Jan **2012**, 100 (1), 13–28 [Online]; http://chess.eecs.berkeley.edu/pubs/843.html

11. Broman, D.; Lee, E. A.; Tripakis, S.; Torngren, M. "Viewpoints, Formalisms, Languages, and Tools for Cyber-physical Systems. In *To appear in Proceedings of the 6th International Workshop on Multi-Paradigm Modeling, October 2012* [Online]; http://chess.eecs.berkeley.edu/pubs/939.html

12. Cardoso, J.; Derler, P.; Eidson, J.; Lee, E. A. Network Latency and Packet Delay Variation in Cyber-physical Systems. In *2011 IEEE 1st International Workshop on Network Science (NSW2011) West Point, NY. IEEE*, June 2011 [Online]; http://chess.eecs.berkeley.edu/pubs/844.html

13. von Hanxleden, R.; Lee, E. A.; Motika, C.; Fuhrmann, H. Multi-view Modeling and Pragmatics in 2020, Position Paper on Designing Complex Cyber-physical Systems. In *Proceedings of the 17th International Monterey Workshop on Development, Operation and Management of Large-Scale Complex IT System, to appear.* LNCS, March 2012 [Online]; http://chess.eecs.berkeley.edu/pubs/905.html

14. Pelino, M.; Hewitt, A. The Forrester Wave: IoT Software Platforms, Q4 2016: The 11 Providers That Matter Most and How They Stack Up, Jul. 2016 [Online]; https://slidex.tips/down-load/res136087-de. Accessed on Mar 31, 2018.

15. Stich, V.; Hoffmann, J.; Heimes, P. Software-definierte Plattformen: Eigenschaften, Integrationsanforder-ungen and Praxiserfahrungen in produzierenden Un-ternehmen. *HMD-Praxis der Wirtschafts informatik* **2018**, *55* (1), 25–43.

16. Crook, S.; Mac Gillivray, C. IDC Market Scape: Worldwide IoT Platforms (Software Vendors): Vendor Assessment, 2017 [Online]; https://www.ge.com/de/sites/www.ge.com.de/files/IDC%20MarketScape_Worldwide%20IoT%20Platforms_Software%20Ven-dors_US42033517%5B1%5D.pdf (accessed on: Mar 30, 2018).

17. IoT Platform—IBM Watson IoT Platform [Online]; https://www.ibm.com/internet-of-things/spotlight/watson-iot-platform (accessed on: Jul 01, 2018).

18. Regazzoni, C. S.; Cavallaro, A.; Wu, Y.; Konrad, J.; Hampapur, A. Video Analytics for Surveillance: Theory and Practice. *IEEE Signal Process. Magazine* Sept **2010**, *16*, 16–17.

CHAPTER 4

Internet of Things: Applications, Vulnerabilities, and the Need for Cyber Resilience

V. KARPAGAM[1*], A. KAYALVIZHI[1], NAVEEN BHARATHI[2], and S. BALAMURUGAN[3]

[1]*Department of Information Technology, Sri Ramakrishna Engineering College, Coimbatore, India Coimbatore, India*

[2]*Software Design/Cloud Computing Consultant, Director, Navster Limited, United Kingdom*

[3]*Founder & Chairman-Albert Einstein Engineering and Research Labs (AEER Labs); Vice Chairman-Renewable Energy Society of India (RESI), India*

*Corresponding author. E-mail: karpagam.vilvanathan@srec.ac.in

ABSTRACT

Industry 4.0 essentially refers to smart manufacturing plants with connected physical devices, which can communicate with each other, exhibiting increased efficiency with access to real time insights and the capability of triggering actions with minimal human involvement. It has become the trending omnipresent term today due to its impact on a wide arena of industries, ranging from healthcare to manufacturing. A broad range of industries is reaping the benefits of Industry 4.0 in terms of increased productivity, efficiency, reduced costs, and real-time decision making. So it has made all this possibilities. It is all owing to the proliferation of connected Internet of Things (IoT) devices and the confluence of human augmentation technologies that include artificial intelligence (AI), robotics, data analytics, cognitive technologies, high computing capabilities, and augmented reality.

4.1 INTRODUCTION

Industry 4.0 essentially refers to smart manufacturing plants with connected physical devices, which can communicate with each other, exhibiting increased efficiency with access to real-time insights and the capability of triggering actions with minimal human involvement. It has become the trending omnipresent term today due to its impact on a wide arena of industries, ranging from healthcare to manufacturing. A broad range of industries is reaping the benefits of Industry 4.0 in terms of increased productivity, efficiency, reduced costs, and real-time decision making. So what has made all this possible? It is all owing to the proliferation of connected Internet of Things (IoT) devices; and the confluence of human augmentation technologies that include Artificial Intelligence (AI), robotics, data analytics, cognitive technologies, high computing capabilities, and augmented reality. Cloud storage is capacious enough to store all the streaming data generated from millions of connected devices. Analytics along with AI and deep learning capabilities help in cost-effective, automated decision making. Robotics enables the transformation of this Information Technology into the corresponding Operations Technology. The advent of 5G is all set to accelerate the momentum of Industry 4.0 with its zero latency and capacity.

Any advancement in technology is always accompanied with its pros and cons–so are the benefits of Industry 4.0. A connected industry is always vulnerable to hacks, as any connected device would be. Though it is impossible to create an environment which is 100% secure, it is the responsibility of the creators of this Industry 4.0 ecosystem to study and identify possible vulnerable points and create a cyber-resilient system. As always, adding security features at the end of the development cycle will not prove to be effective in terms of security as well as cost. The need, therefore, is to build cyber resilient systems, with a Secure Software Development Life Cycle (SSDLC) that integrates security into the development cycle.

This chapter aims to discuss the benefits Industrial IoT. Three specific applications have been considered to discuss the features, benefits, potential risks, and vulnerable points in an application. The chapter also summarizes the need for various security services that have to be considered in these systems and possible solutions have been suggested.

BENEFITS OF INDUSTRY 4.0:

- **Increased efficiency and productivity:** Continuous or regular monitoring along with automated analysis and decision making has increased the efficiency and productivity of manufacturing plants.
- **Prediction:** Systems have gained the knack to foresee and reckon solutions to problems.
- **Industrial safety:** Automation can be employed in places where currently human intervention is required in hostile working environments.
- **Increased revenues and greater profit margins**
- **Consumer engagement:** The data available give better insights on customer's needs.
- **Multiple vendor support**

4.2 NEED FOR CYBER RESILIENCE

Due to the numerous benefits, industries are quickly transforming to Industry 4.0. However, a better focus on the security of the connected systems has become the need of the hour to ensure that the connected devices are not vulnerable to malicious activities. Figure 4.1 depicts the escalation in the number of cyber threats along with each industrial revolution.

4.3 APPLICATIONS OF INDUSTRY AUTOMATION

Three applications of Industrial Automation and possible security threats[1] are discussed in this section:

I. Autonomous and connected vehicles
II. Digital/connected factory
III. Logistics and supply chain management

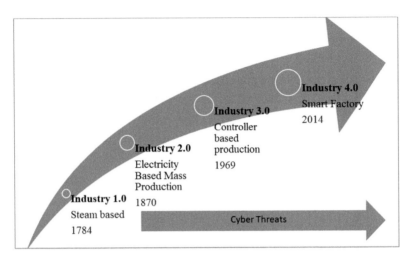

FIGURE 4.1   Escalation in the number of cyber threats along with each industrial revolution

Source: Adapted from https://www2.deloitte.com/us/en/insights/focus/industry-4-0/cybersecurity-managing-risk-in-age-of-connected-production.html

I.   *Autonomous and connected vehicles*

Industry 4.0 refers not only to the manufacturing sector, but covers a wide range of sectors. Autonomous vehicles are a good application of Automation. Tesla car is a popular example of driverless cars sporting features such as Wi-Fi, Bluetooth, and built-in garage door. Autonomous vehicles deploy a number of sensors like LiDAR, Radars, Imaging from Cameras, and sensors for GPS Geolocation information. Numerous sensors are also employed for information on attitude (angle) and acceleration and a camera.

Autonomous cars are expected to include features such as:

- Sensing the context of the environment
- Identify the correct paths for navigation
- Detect obstacles and act suitably
- Interpret road signs
- Drive in low visibility conditions
- Limit the vehicle's movement to a particular lane
- Apply emergency braking
- Localization and mapping capabilities

Threats: Now it can be left to anyone's imagination about the amount of disaster that can happen if someone can hack into this connected autonomous vehicle and take complete control of it. The threat[2,4] in the case of hacking an autonomous vehicle would not be limited to the vehicle alone, but could have devastating effects on the environment around it.

- The autonomous vehicle could be fatal to both passengers and others if it were hacked inappropriately. What if an unauthorized third party is able to:
  - apply brakes and stop the fast moving vehicle.
  - accelerate when it is necessary to apply brakes.
  - hack the sensors to communicate wrong information about the environment.
  - open and close doors at the will of the malignant user only.
  - have unauthorized access to location information (Fig. 4.2).
  - in case of autonomous drones (Fig. 4.3), what if pictures of private locations are accessed.

- In case of industry environments, autonomous vehicles find a wide range of applications, like logistics, indoor transportation of goods; used in areas where human entry might be dangerous, etc. Autonomous vehicles in industry environments operate with less stringent safety requirements in comparison with driverless cars with passengers.

FIGURE 4.2 IoT Tesla car.

FIGURE 4.3   Autonomous drones.

- A user with malicious intent can cause significant damage to the whole system, if he/she is able to interfere with the GPS information sent to the vehicle.
- The vehicle is dependent on the information sent from all the sensors, so making this data unavailable could be significant threat.
- Key or password attacks to gain access to the vehicles operations may be possible.
- Denial of service attacks can be launched by converting nearby vehicles connected with Wi-Fi into botnets.
- Smart cars have information about all functions and location information stored in its database. Unauthorized access to information in this database may turn out to be a threat to privacy.

II. Digital/connected factory

Factories are now capable of adapting to real-time workflows – all made possible with machine to machine (M2M) communication and human to machine communication. Industry 4.0 brings together people, processes and products enabling consumer engagement, continuous value delivery to the customers and improved efficiency in all stages of the production lifecycle.

Connecting all of the parts of the manufacturing process has simplified the development and testing of applications on different platforms is. The monitoring of the performance and utilization of all devices/applications, preparing schedule and workload of employees, better utilization of the existing infrastructure and predicting the investment needs for business growth has all become easy with the connected factories and all data available for insights (Fig. 4.4).

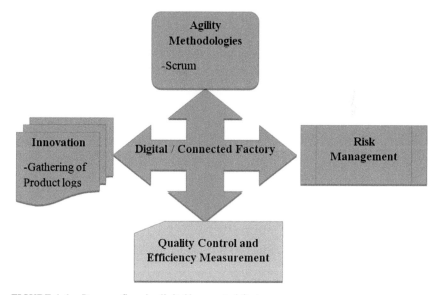

FIGURE 4.4 Process flow in digital/connected factory.

The connected devices in an Industry 4.0 factory are capable of integrating applications, data and processes. Storage options include the device itself, local storage in the premises of the factory and cloud storage. In case of a disaster, it is possible to recover an entire data center in just hours today.

The monitoring system continuously polls the data streamed from the sensor or devices, and tries to predict or detect a failure in the system. Machine learning techniques are deployed to foresee problems with equipment and help early replacement, preventing long downtime of the system. Information from the sensors is also used to locate the faulty devices. Predictive maintenance helps in lowering the costs in a connected factory.

Threats: What if a hacker:

- launches a denial of service attack on the centralized data center that could make the entire factory inoperative

- ables to issue commands and controls the operations of devices available at the factory
- intercepts data in transit from the devices to the data centers or from the command and control systems to the devices.

III. Supply chain management

Automation of supply chain management[8] in Industry 4.0 involves the analysis of the huge volume of digitized data which is available to automate order processing and ensuring availability of data which is critical for continuous production. The advantages of applying the confluence of technologies of Industry 4.0 to supply chain management have resulted in the following benefits:

- Reduces inventory and capital requirements
- Demand forecasting
- Dynamic pricing
- Manage the flow of materials and resources in a more efficient way
- Meeting the needs of the customer more appropriately

The risks of hacking a supply chain may have many implications like financial loss, low quality products, loss of reputation, safety concerns, and risks due to fake suppliers. The kind of attacks may include interception of information, breach of confidentiality of information like competitive pricing, or intellectual properties of the vendors. Man-in-the-middle attacks may be launched to intercept, destroy, or fabricate information.

Blockchain[6] appears to be a panacea to trace the flow of goods and money in a supply chain. The support for multiple vendors is a major advantage of Industry 4.0. Block chain will help to identify any vendor in the supply chain and help to trace the source of supply for any goods.

4.4   SECURITY MEASURES

The implications for a system which is not cyber resilient may be loss in terms of production downtime, equipment damage, loss of life, monetary loss, litigation expenses, brand damage, threat to intellectual property and privacy. There is a need therefore to build cyber resilient systems (Fig. 4.5).

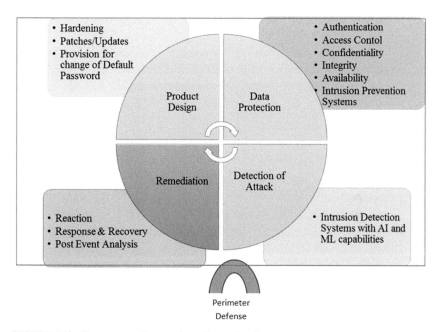

FIGURE 4.5 Focus areas for attack surface reduction.

The security measures for attack surface reduction and creation of cyber-resilient systems in Industry 4.0 is discussed over here from four focus areas:

1. Product Design
2. Data Protection
3. Threat/Attack Detection
4. Remediation/Response

- *Product Design:*
  - For any particular functionality of a device only services which are absolutely necessary should be visible or accessible.
  - Devices should be inherently secure or most of the security features should be hardened in the devices themselves. This is especially required due to the skill shortage for personnel in cyber security.
  - Choice of COTS–Off the shelf components should be made with care and verified to see if they provide the required security features.

- Devices should be developed following the (S-SDLC) Secure Software Development Life Cycle.
- The product manufacturers should be responsible for security patches and updates for the devices, because unknown threats may arise or vulnerabilities may be found throughout the life cycle of the product.

- *Data Protection*
  - Most of the industries today are using cloud storage[7] for their data and data from any command and control centre are also sent to the devices over Wi-Fi or network communications. These data in transit are susceptible to attacks of confidentiality breach, masquerading, interception, fabrication, modification, and also destruction of the data. The confidentiality attacks can be handled through encryption services. The integrity and authentication of data can be determined through message digest, hashing algorithms, and digital signatures.

- *Threat/Attack Detection*
  - With the number of devices in a connected factory on the rise, it is difficult to just deploy the traditional methods of Intrusion Detection and Prevention systems. The use of AI[5] and Machine learning solutions for Intrusion Prevention and Intrusion Detection Systems is the need for the hour to enhance the security of these connected systems.
  - Change of default passwords is a major requirement for security in a system with numerous connected devices. In most of the reported cases of Distributed Denial of Service attacks (DDoS),[9] devices with default passwords have been converted to botnets to launch DDoS attacks.

- *Remediation*
  The remediation or response to any attack is important. Any attack or vulnerabilities discovered have to be patched up and carefully and consequently considered as a feedback for secure design of the device and similar devices.

Recommendations for Creation of a Secure Industry 4.0 Ecosystem

1. Network segmentation
2. Protection of intermediary systems
3. Hardware security–hardware authentication and integrity
4. Access control
5. Encryption and tokenization of data
6. Perimeter protection
7. Blockchain
8. Stringent security policies enforced by managements

KEYWORDS

- **Internet of things**
- **Industry 4.0**
- **robotics**
- **data analytics**
- **cloud computing**

REFERENCES

1. http://www.sio2corp.com/cyber-security-applications/
2. https://www.globalsign.com/en-in/blog/cybersecurity-trends-and-challenges-2018/
3. https://blog.netwrix.com/2018/05/15/top-10-most-common-types-of-cyber-attacks/
4. https://link.springer.com/chapter/10.1007/978-3-642-32021-7_11
5. https://www.jkoolcloud.com/artificial-intelligence-expansion-adoption-value-business-intelligence/
6. http://blockchain-revolution.com/
7. https://serverless-stack.com/chapters/what-is-serverless.html
8. https://www.supplychain247.com/article/identifying_cyber_vulnerabilities_in_manufacturing_supply_chains
9. https://www2.deloitte.com/content/dam/insights/us/articles/3749_Industry40_cybersecurity/DUP_Industry4-0_cybersecurity.pdf

CHAPTER 5

Medical Cyber-Physical Systems Security

R. REKHA

Department of Information Technology, PSG College of Technology, Coimbatore 641004, India

*Corresponding author. E-mail: rekha.psgtech@gmail.com

ABSTRACT

People with chronic illnesses such as cardiovascular disease, stroke, and diabetes can be continuously monitored with the help of Remote Patient Monitoring (RPM) systems. RPM is a form of telehealth that uses small sensor nodes to collect medical data from remote patients and telecommunications to deliver the collected data to healthcare systems. The sensor nodes that are wearable or implanted in a patient's body constitute the medical Body Area Sensor Network (BASN). These sensors transmit the collected data wirelessly to the master node located on the patient's body. The master node forward the data to a centralized repository, where the diagnosis of arrhythmia class is performed and alerts are sent to healthcare providers. This chapter focuses on securing the inter-sensor communication in Wireless BASN (WBASN) by agreeing upon a symmetric cryptographic key using electrocardiogram signals. An improved fuzzy vault scheme is proposed to minimize false acceptance rate (FAR), false rejection rate (FRR) and to overcome correlation attack.

5.1 REMOTE PATIENT MONITORING SYSTEM

Remote patient monitoring (RPM) system helps patients with chronic diseases to remain associated with their clinical care group without massive wires or difficult to operate gadgets. The system wirelessly transmits the patient's vital signs to the remote server. Figure 5.1 shows the architecture of healthcare monitoring systems that collect medical data using body area sensors and communicate to remote healthcare providers through an aggregation unit like personal computer or mobile phone.

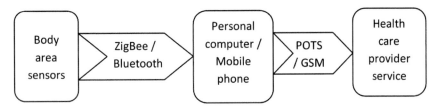

FIGURE 5.1   Architecture of healthcare monitoring system.

Due to the availability of small, lightweight and ultra-low power sensors, continuous monitoring of a person's health is possible and it helps in the early detection of potential illnesses. Heart rate, respiration rate, electrocardiogram (ECG) recording, and highly accurate positional and activity information are updated periodically with the help of body area sensors. Continuous update of vital health information helps clinicians to provide the best care to their patients with chronic diseases such as heart failure, stroke, and diabetes.

The main components in the design of RPM frameworks are, (1) The sensors on a patient's body that communicate wirelessly framing Wireless Body Area Sensor Network (WBASN), (2) The master node at the patient's site that serves as an interface between the sensors and the centralized data repository and provides local data storage and, (3) Centralized repository to accumulate the data obtained from the master node and other diagnostic applications and (4) Diagnostic software that produces alerts based on the analysis made with collected medical data.

5.2   WIRELESS BODY AREA SENSOR NETWORK

Recent technological advances in wireless networking guarantee a new generation of wireless sensor network suitable for on-body/in-body sensor networked systems called WBASN. A number of tiny wireless sensors that are strategically placed on/in a patient's body and a master node on the same patient's body are connected with each other to create a WBASN.

WBASN can monitor vital signs, thereby providing real-time feedback through the continuous monitoring of chronic conditions. Figure 5.2 shows a simple example of a WBASN. Medical information collected by the sensors in a WBASN is sent to a highly capable master node for data fusion. From the master node, the medical data are sent to a centralized data repository for further processing.

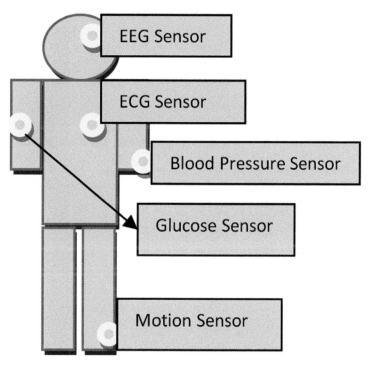

FIGURE 5.2   A simple wireless body area sensor network.

5.3   SECURITY CHALLENGES TO WIRELESS BODY AREA SENSOR NETWORKS

As per the Health Insurance Portability and Accountability Act (HIPAA), medical applications should ensure the confidentiality of the patient's medical data. Sensor nodes can communicate without wires. There-fore, there might be risks of passive attacks, like observing information exchanged or eavesdropping of medical data. This has an effect on the privacy of the patient's health information. Active attacks like modification of information communicated to the recipient may bring about life-threatening circumstances. Hence, preserving confidentiality, integrity, and authenticity of sensor nodes are very essential.

Sensor nodes work with low power, limited memory and have low-computational capability. Therefore, usage of a separate random number generator for cryptographic key generation is not feasible. Likewise, complex security mechanisms cannot be used to secure the wireless transmission of medical information between the sensors of a patient's body. A solution to this is to use the physiological signals captured by sensor nodes to resolve the security issues in WBASN.

Secure communication in WBASN should be achieved during,

1. Communication between sensor nodes of same WBASN.
2. Communication between the master node and base station.
3. Communication between the base station and medical server.

The objective of this work is to develop a key agreement scheme to secure inter-sensor communication. Secure communication between the master node and the base station is achieved using MD5 and DES. Secure communication between the base station and the medical server is achieved using steganography.

Several encryption and key exchange algorithms are designed to secure the communication of WBASN. But, the encryption and key management schemes use keys that are exchanged between the transmitter and the receiver at one time or the other, or, use broadcasted keys in refreshment schedules. In such a scenario, if the key exchange communication is somehow intercepted, the attacker could easily get access to all the subsequent frames exchanged between the communicating entities. Thus, even though the system seems to be secure, it is still vulnerable.

This chapter illustrates the agreement of cryptographic keys between communicating sensors by utilizing ECG, the physiological signal. Using the proposed key agreement scheme, secure inter-sensor communication can be achieved even when new sensor nodes are added.

5.4 PHYSIOLOGICAL SIGNAL BASED KEY AGREEMENT

Usage of ECG for secure key exchange among sensors of the same patient's body avoids the need for a separate mechanism for symmetric key agreement. ECG signals measured at different parts of the same body may be highly correlated but need not be exactly the same. In order to uncover the key that is hidden and transmitted by the sender, it is essential for the receiver to have the exact features extracted from its own ECG signal. Hence, it is advantageous to use fuzzy logic which can accommodate slight variations in the ECG signal features that are extracted at different sites of the same body.

Hatzinakos and Bui (2008) proposed a key distribution method through multipoint fuzzy key management and key fusion. Keys were generated using ECG signals and they proved to provide good randomness similar to the key generated partially with external random number source. Venkatasubramanian et al. (2010) proposed a fuzzy vault scheme to overcome feature mismatch at the sender and receiver ends. Key agreement among biosensors of the same person can be achieved by using a shorter segment of the ECG signal captured and processed in frequency domain. The strength of the method relies on polynomial reconstruction and the random chaff points added. Usage of chaff points was found to increase the communication overhead and memory requirement. The drawback of the system was the high probability for the adversary to guess the genuine points.

Zhang et al. (2012) proposed a key agreement using ECG and the Improved Jules Sudan method. ECG signals were used to generate keys. The efficiency of this method relied on the optimal selection of the different tolerance parameter. Chaff points were not required for the key agreement. Hu et al. (2013) proposed a key agreement scheme using overlapping ECG signal features. Along with genuine features, chaff was also added. The nodes that belong to the same person will have some features common with that of coffer received. The matching indexes were transmitted back to the sender and the key was generated using the matching features. The drawback was that the communication overhead increased with the coffer size. FAR was decreased when the threshold for the number of matching features was set more than 10 for a coffer size of around 2000.

Literature survey on systems that use biometrics for key generation and key agreement is tabulated in Table 5.1.

5.5 IMPROVED FUZZY VAULT KEY AGREEMENT SCHEME

The proposed improved fuzzy vault key agreement scheme hides a secret (H) in a construct called a vault using a set of values from sender (S).

TABLE 5.1 Literature Survey on Biometrics for Key Generation and Key Agreement

Study (year)	Method proposed	Research finding	Inference
Cherukuri et al. (2003)	Generation and use of pseudo random numbers from the properties of human body at different sites.	Randomness is high when the sequence is derived from multiple biometrics simultaneously.	Biometric keys have sufficient randomness.
Poon et al. (2006)	Timing information of heartbeat can be an excellent biometric characteristic to secure BASN.	Inter Pulse Interval (IPI) have high randomness and the error rate increases when there is asynchrony between IPIs at the sender and the receiver sides.	IPI can be used to secure BASN.
Hatzinakos and Bui (2008)	Key distribution through multipoint fuzzy key management and key fusion.	The number of bits transmitted over the channel for key distribution is less.	Keys generated using ECG show good randomness.
Bao et al. (2008)	IPI to generate entity identifiers.	Generated EIs have sufficient randomness.	Computational complexity is slightly increased in each sensor node.
Zhang et al. (2009)	Keys are generated using static and dynamic biometrics.	Entropy of encryption keys range from 0.662 to 1.	Distinctive keys can be generated from static and dynamic biometric traits.
Zhang et al. (2010)	Key generation from segments of cardiac signal and derivative of the signal.	Entropy of the key generated was above 0.95 and the Hamming distance of the pair of keys generated was above 50 bits.	Sufficient randomness and distinctiveness can be achieved from the generated key.
Venkatasubramanian et al. (2010)	Fuzzy Vault scheme to overcome feature mismatch at sender and receiver sides.	Security of secret key hiding is based on the number of chaff points included.	High probability for the adversary to guess the genuine points.

TABLE 5.1 *(Continued)*

Study (year)	Method proposed	Research finding	Inference
Xu et al. (2011)	ECG delineation based key generation.	Prior sharing of secrets among IMDs and guardian is not required. Able to generate exactly symmetric keys from ECG.	Time overhead to obtain exactly matched keys may increase due to block discard when there is a mismatch.
Zhang et al. (2012)	Key agreement using ECG and Improved Jules Sudan method.	ECG signals are used to generate keys. The efficiency of this method relies on the optimal selection of the difference tolerance parameter.	Prior key distribution is not required. Chaff points are not required.
Hu et al. (2013)	Key generation scheme from overlapping ECG signal features.	Communication overhead increases with coffer size. FAR decreases when the threshold is set high for the number of matching features.	Resistance against brute force attack. Key pre distribution is not required.

This vault can be unlocked only by the receiver (R) which has significant number of values in common with the sender. The purpose of the key agreement scheme proposed is to establish the secure key exchange among sensors in the same body.

Correlation attack is the possibility for an attacker to find out the secret encapsulated in the fuzzy vault by correlating two or more previously transmitted vaults from a sender (Kolmatov and Yanikoglu, 2008). The proposed scheme achieves secure key agreement and overcomes correlation attack through the subsequent phases:

1. Authentication of communicating sensor nodes through verification of Hamming distance between the random binary sequences generated from the Inter Pulse Intervals (IPIs) collected by those sensors.
2. Key agreement through fuzzy vault scheme with template creation, linear transformation and chaff generation.

5.5.1 RANDOM SEQUENCE GENERATION FROM IPI

Random sequences are generated using the IPI of ECG signals. Experimental outcomes exhibited that the generated binary sequence is highly random when four bits are extracted from each IPI. Therefore, a 128-bit random binary sequence is generated by concatenating the extracted bits from 32 IPIs. The random sequence is given by, $R = (b_0|b_1|b_2|.....b_{31})$, where b_0, b_1....are binary sequence extracted from each IPI. Initial authentication of communicating sensors is performed by verifying the Hamming distance of the binary sequences. By experimentation, it is found that when the Hamming distance between binary sequences generated by two sensors is greater than 14, then the sensors does not belong to the same patient. Therefore, authentication failed and further communication is prohibited.

5.5.2 FEATURE EXTRACTION

Feature extraction techniques transform raw pre-processed signals into more informative, non-redundant features. Sender and receiver nodes of WBASN collect ECG signals simultaneously for duration of 4s. Fast

Fourier Transform (FFT) is used to extract the features from the continuously monitored ECG signal. Fourier Transform decomposes a signal into complex exponential functions and expresses it in terms of different frequencies of the waves that make up the signal. Any deviations in the regular shape of the normal ECG beat can be easily visualized by analyzing the frequency content of the ECG segment. This work utilizes FFT for feature extraction. FFT is a discrete Fourier transform (DFT) algorithm that reduces the number of computations needed for "N" points from O (N^2) to O (N log N) operations. The formula for computing the DFT of a given ECG sequence is given by equation (5.1),

$$X_k = \sum_{n=0}^{N-1} x_n e^{-i2\pi k \frac{n}{N}} \quad k = 0, \ldots, N-1 \tag{5.1}$$

where x_n is the input ECG sequence and X_k is the Fourier transformed sequence of ECG samples. The absolute phase is retained by this transform and it provides more information than the relative phase. This characteristic is an advantage compared to wavelet transform. Collected ECG signals are sampled at a sampling rate of 60 Hz. FFT is performed over the samples after dividing it into windows. Peak-detection function is applied over the FFT coefficients and features are generated by concatenation of the peak index vales and its corresponding FFT coefficient value. The feature vector computed by the sender node is given by $F_s = \{f_s^1, f_s^2, f_s^3 \ldots f_s^m\}$ and the feature vector computed by receiver node are given by $F_r = \{f_r^1, f_r^2, f_r^3 \ldots f_r^m\}$, where "m" is the size of the feature vector.

The metrics used to evaluate the work are defined as follows:

- FAR is a measure of the likelihood that the system will falsely accept an access by an unauthorized user.

$$FAR = \frac{\text{Number of matching features between patients A and B}}{\text{Total number of features transmitted by patient A}}$$

- FRR is a measure of the likelihood that the system will falsely reject an access by an authorized user.

$$FRR = \frac{\text{Number of mismatch in features between sensor nodes of same patient A}}{\text{Total number of features transmitted by patient A}}$$

- Half Total Error Rate (HTER) = $\dfrac{FAR + FRR}{2}$

The block diagram of secure key agreement scheme is shown in Figure 5.3.

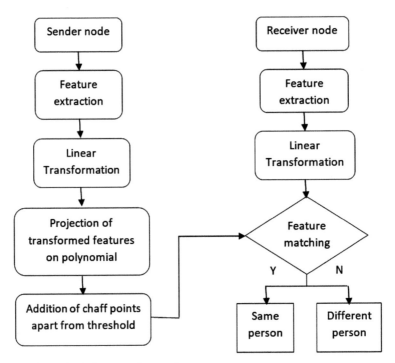

FIGURE 5.3   Block diagram of secure key agreement.

5.5.3   *TEMPLATE CREATION*

Even though the ECG signals are collected from the same patient; at the same time instance, but from different locations of the patient's body, there might be a slight deviation in the captured signal. If the chaff points added at the sender side are very nearer to the valid point generated at the sender, then there is a chance for the receiver feature to match with chaff present in the vault leading to an increase in FRR. Hence, a template creation will avoid such issues.

Feature template is created both at the sender and receiver by capturing ECG signals for a duration of 30 s and repeating the same steps of feature extraction. Out of 256 FFT coefficients, the first 128 coefficients are extracted from each window. The arithmetic mean of coefficients for constructing the template is computed as $[fT_{w11} + fT_{w21} + \ldots + fT_{wn1}]/n \ldots \ldots$ $[fT_{w1m} + fT_{w2m} + \ldots + fT_{wnm}]/n$, where "n" is the number of windows and "m" is the number of coefficients extracted from each window. The sender and receiver nodes use their own captured signal to compute the average of coefficients in the same manner using $[fS_{w11} + fS_{w21} + _{\ldots} fS_{wn1}]/n \ldots$ $[fS_{wm1} + fS_{wm2} + _{\ldots} fS_{wnm}]/n$ and $[fR_{w11} + fR_{w21} + _{\ldots} fR_{wn1}]/n \ldots [fR_{wm1} + fR_{wm2} + _{\ldots} fR_{wnm}]/n$. Here, fT denotes the average value of the template obtained from 30 s ECG signal, fS denotes the average value obtained from 4 s ECG signal of the sender and fR denotes the average value obtained from 4 s ECG signal of the receiver. Using the template and the corresponding feature vectors, the maximum difference (Max_diff) and the minimum difference (Min_diff) are calculated. Interval value is found by, interval = (Max_diff − Min_diff). The template is varied periodically to ensure freshness.

5.5.4   LINEAR TRANSFORMATION

Extracted features are linearly transformed to overcome the possibility of correlation attack in the transmitted fuzzy vaults.

5.5.5   FUZZY VAULT CREATION

The sender node generates a dth order polynomial of the form $p(x) = c_v x^v + c_{v-1} x^{v-1} + \ldots + c_0 x^0$. The steps followed to project the features on polynomial at sender are given by:

1. Generate the biometric key from the physiological signal by extracting four bits from each IPI.
2. Assume that "k" denotes the biometric key (secret) to be hidden using the set $F_s = \{f_s^1, f_s^2, f_s^3, \ldots f_s^m\}$ available with the sender.
3. Secret "k" is embedded on the coefficients of the single variable polynomial "p" selected with degree "d".
4. Concatenation of the polynomial coefficients gives the secret key to be shared by the sender and the receiver node, i.e., secret key = $(c_0, c_1, c_2, \ldots c_v)$. The length of the key is fixed to be 128 bits.
5. Considering the elements of set F_s as distinct coordinate values of the polynomial, evaluate the polynomial p on the elements of F_s and compute the set $V = \{F_{si}, p(F_{si})\}$, where i = 1 to n.

6. Create a number of random chaff points "F" that do not lie on the polynomial p. (i.e.) $F = \{f_j, t_j\}$, where f_j does not belong to set F_s, $t_j \neq p(t_j)$ and j = 1 to m.
7. The entire set of point, that is valid points and chaff points constitute the vault "S" that is sent to the receiver.

Chaff Point Generation. The security of the fuzzy vault scheme depends on the number of additional chaff points added along with the valid points. These chaff points conceal the valid points from the hackers. Moreover, the chaff points are selected in such a way that they also lie in the same range as that of valid points. For example, if F_s can have any value in the range 0–256, then F can also be within the same range. The interval value calculated using template is checked and random chaff points are placed apart from that interval. Placing chaff points apart from template avoids the matching of valid features from chaff points.

5.5.6   RECEIVER

1. Set $F_r = \{f_r^1, f_r^2, f_r^3, ... f_r^m\}$ is the feature vector generated at the receiver end and it substantially matches with set F_s.
2. Compute a set M, with the help of the received vault and elements of F_r, such that $M = \{(b, c)|(b, c) \in S \text{ and } b \in F_r\}$.
3. With the set M, the receiver then tries to compute the actual polynomial generated at the sender.

The hardness of this scheme relies upon the polynomial reconstruction problem. Hence, in order to reconstruct a polynomial of degree "d", a minimum of "d + 1" matching elements should be present between the sender and receiver. To a hacker, it is very difficult to guess the valid points from the vault received and hence, it is difficult to reconstruct the actual polynomial.

5.5.7   SECURITY ANALYSIS

The security of the proposed scheme is based on the difficulty of polynomial reconstruction. Hiding the legitimate feature points in a larger number of the chaff points, whose values are in the same range as that of legitimate points, makes the job of identifying the legitimate points very difficult.

Figure 5.4 shows the strength of the vault for different polynomial orders and when varying the number of chaff points.

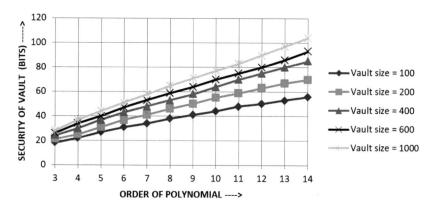

FIGURE 5.4   Vault strength for different polynomial orders and vault sizes.

Increasing the number of chaff points increases the security provided by the vault. When the order of the polynomial is increased, the number of common features between sender and receiver required to reconstruct the polynomial is increased thereby increasing the security. For example, if a hacker captures the transmitted vault constructed using the polynomial of order $= 8$ with vault size $= 200$, then the hacker has to try 5.5×10^{13} combinations of features to reconstruct the polynomial. For easy interpretation, the security of the vault is represented in Figure 5.3 in terms of the count of binary bits representing the total combination required.

5.5.8   RESULT ANALYSIS

Fifteen patients are randomly selected from the MIT-BIH Arrhythmia database (patient identifier: 100, 102, 103, 105, 106, 107, 108, 109, 112, 113, 114, 115, 116, 117, and 118). ECG signal is collected from each ECG record for duration of 4 s. The original ECG signal is sampled and 256 points FFT is applied and the peaks are detected. After detecting the peaks, peak index and peak values are concatenated and taken as features. For example, if peak index $= 1$ and peak value of that index $= 14$, then the generated feature is $f_s^1 = 114$. The generated features are evaluated on the polynomial selected and the vault is created by the sender. The main drawback of the existing system is that sometimes the receiver's features may match with the chaff points and lead to increase in FRR. The proposed methodology overcomes this drawback with the help of template creation.

The problem with the existing system is that the receiver may mistakenly consider some chaff points in the received vault as matching features. Vault constructed using the proposed system with 200 chaff points placed according to the interval value calculated with the help of template reduces the mismatch by the receiver. Since the mismatch of chaff points as valid points is reduced, FAR gets minimized. Once the vault is received, the receiver uses its own extracted features and tries to find the common features from the vault.

Figures 5.5 and 5.6 show the average FRR and average FAR for vaults constructed using various orders of the polynomial. FRR represents the number of times the common features between sensor nodes of the same patient falls below the value "d + 1", where "d" is the polynomial order.

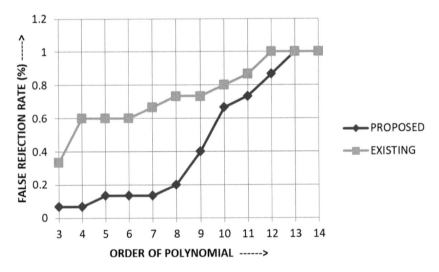

FIGURE 5.5 Comparison of average false rejection rate.

FRR increases as the order of the polynomial increases. This is due to the fact that if more common elements of the feature are needed, there is more probability for the system to reject the two feature sets that come from the same patient. Compared to the existing system, the proposed method minimizes the FRR with the help of the template created.

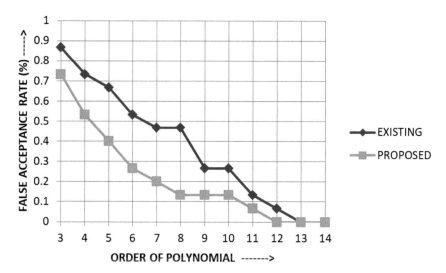

FIGURE 5.6 Comparison of average false acceptance rate.

The number of times the common features between two different patients fall above the value of "d + 1" characterizes FAR. Figure 5.7 demonstrates the percentage of FAR against FRR for the proposed system.

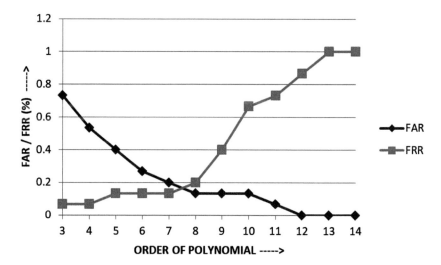

FIGURE 5.7 False acceptance versus false rejection rates for the proposed system.

The FAR decreases with increase in polynomial order and is shown in Figure 5.6. The polynomial order to construct the vault has to be selected in such a way that both FAR and FRR are minimum.

From experimental results, it is found that polynomial order = 8 minimize both false acceptance and FRR. When the order of polynomial is increased beyond eight, the FRR is increased. FAR is increased when the polynomial order is decreased below eight.

The advantage of the proposed fuzzy vault system is that it tolerates the mismatches in the features produced at the valid sender and receiver ends with the help of template and still permits symmetric key agreement without any pre-deployment. The drawback of this fuzzy vault system is the likelihood of several attacks such as collusion/correlation attack, cross-matching attack, and record multiplicity attack, when generated vaults are wirelessly transmitted by communicating sensor. The proposed scheme overcomes correlation attack and achieves key agreement between communicating sensors of a patient in a secure manner.

5.5.9 SECURITY ANALYSIS

The initial authentication of the sender node with the receiver node is achieved by verifying the Hamming distance. The vault is received by the receiver node only when the communicating sensor node proves authentication with Hamming distance less than 14. Experiments show that the extent of correlation between vaults formed without transformation is on average six valid features. Hence, by relating the past vaults transmitted, an attacker can easily unlock a transmitted vault that uses a polynomial of degree five. However in the proposed framework, it is found that the number of correlating features is only two on average. It is practically not possible to construct the polynomial through brute force attack without knowing the valid feature points.

Chaff points are generated in the same space as that of the valid points. Therefore, it is hard to reconstruct the polynomial by guessing the valid points. In addition, legitimate features (valid points) are extracted from non-stationary ECG signals. Therefore, the features extracted from the time varying ECG signal accompanied by the transformation technique create vaults that vary for each transmission.

5.5.10 *PERFORMANCE ANALYSIS IN TERMS OF FAR/FRR*

FAR and FRR should be minimized for a better operating framework. The FRR decreases and FAR increases when the number of valid points projected on the polynomial at the sender end is increased from 15 to 30. Experimental results demonstrate that the FAR decreases, when the vault is constructed using transformed features. Moreover, chaff points are positioned in a manner that they are away from the valid points by a threshold computed utilizing template. Hence, FRR is decreased and the performance of the proposed framework is enhanced.

5.6 CONCLUSION

A key agreement scheme that facilitates secure symmetric key exchange is developed. Security analysis demonstrates that the proposed fuzzy vault scheme overcomes the correlation attack. The time-varying signal like ECG is used for key agreement; hence, the proposed scheme ensures the freshness of the secret key that is used for subsequent encryption process. FAR and FRR are reduced by 25% in the proposed scheme. The performance of the proposed fuzzy scheme can be further improved by extracting more unique features for individuals and by adopting the optimal vault size. Future work aims at securing the communication from the base station to the medical server. ECG signal collected using biosensors will be securely sent to the master node in a patient's body with the help of cryptographic algorithms that use physiology-based key agreement schemes. From the master node, the collected data are sent to the aggregation unit where identification of the arrhythmia class is performed.

KEYWORDS

- wireless sensor network
- RPM
- BASN
- wearable device
- security

REFERENCES

1. Bao, S.-D.; Poon, C. C. Y.; Zhang, Y.-T.; Shen, L.-F. Using the Timing Information of Heartbeats as an Entity Identifier to Secure Body Sensor Network. *IEEE Trans. Info. Technol. Biomed.* **2008,** *12* (6), 772–779.
2. Cherukuri, S.; Venkatasubramanian, K. K.; Gupta, S. K. S. Biosec: A Biometric Based Approach for Securing Communication in Wireless Networks of Biosensors Implanted in the Human Body. *Proc. Int. Conf. Parall. Process. Workshops* **2003,** *2003* January, 432–439.
3. Hatzinakos, D.; Bui, F. M. Biometric Methods for Secure Communications in Body Sensor Networks: Resource-efficient Key Management and Signal-level Data Scrambling. *EURASIP J. Adv. Sign. Process.* **2008,** 529879 (2007). https://doi.org/10.1155/2008/529879.
4. Hu, C.; Cheng, X.; Zhang, F.; Wu, D.; Liao, X.; Chen, D. OPFKA: Secure and Efficient Ordered-physiological-feature-based Key Agreement for Wireless Body Area Networks. *Proc.—IEEE INFOCOM* **2013,** 2274–2282.
5. Kholmatov, A.; Yanikoglu, B. Realization of Correlation Attack against the Fuzzy Vault Scheme, 2008.
6. Poon, C. C. Y.; Zhang, Y. T.; Bao, S. D. A Novel Biometrics Method to Secure Wireless Body Area Sensor Networks for Telemedicine and M-health. *IEEE Commun. Magaz.* **2006,** *44* (4), 73–81.
7. Venkatasubramanian, K. K.; Banerjee, A.; Gupta, S. K. S. PSKA: Usable and Secure Key Agreement Scheme for Body Area Networks. *IEEE Trans. Info. Technol. Biomed.* **2010,** 14 (1), 60–68.
8. Xu, F.; Qin, Z.; Tan, C. C.; Wang, B.; Li, Q. IMDGuard: Securing Implantable Medical Devices with the External Wearable Guardian. *Proc.—IEEE INFOCOM* **2011,** 1862–1870, doi: 10.1109/INFCOM.2011.5934987.
9. Zhang, G. H.; Poon, C. C. Y.; Zhang, Y. T. A Biometrics Based Security Solution for Encryption and Authentication in Tele-healthcare Systems. *2nd International Symposium on Applied Sciences in Biomedical and Communication Technologies*; ISABEL, 2009.
10. Zhang, G. H.; Poon, C. C. Y.; Zhang, Y. T. A Fast Key Generation Method Based on Dynamic Biometrics to Secure Wireless Body Sensor Networks for p-health, 2010; pp 2034–2036.
11. Zhang, G. H.; Poon, C. C. Y.; Zhang, Y. T. Analysis of Using Interpulse Intervals to Generate 128-bit Biometric Random Binary Sequences for Securing Wireless Body Sensor Networks. *IEEE Trans. Info. Technol. Biomed.* **2012,** *16* (1), 176–182.
12. Zhang, Z.; Wang, H.; Vasilakos, A. V.; Fang, H. ECG-Cryptography and Authentication in Body Area Networks. *IEEE Trans. Info. Technol. Biomed.* **2012,** *16* (6), 1070–1078.

CHAPTER 6

Secure Data Aggregation Using Cyber-Physical Systems for Environment Monitoring

M. K. SANDHYA[1*], K. MURUGAN[2], and S. PRASIDH[3]

[1]*Department of CSE, Meenakshi Sundararajan Engineering College, Chennai, India*

[2]*Ramanujan Computing Centre, Anna University, Chennai, India*

[3]*Director, Product Management, Bitglass, Campbell, USA*

Corresponding author. E-mail: mksans@gmail.com

ABSTRACT

Cyber physical system (CPS) can monitor the environmental conditions using wireless sensor networks (WSNs). WSNs comprise of large number of sensor nodes that are densely deployed in a geographical area where specific events have to be monitored or detected. These sensor nodes are low power devices with limitations on computing capability and storage capacity. These sensor nodes sense the occurrence of events and transmit the sensed data to the base station. Sensor nodes produce huge amount of data which needs to be aggregated for effective summarization. False data injected by the compromised sensor nodes will distort the summarized results. Most of the existing security mechanisms in WSNs employ trusted aggregator nodes for eliminating false data injected by the compromised sensor nodes exploiting the spatiotemporal correlation. They do not address the issue of compromised aggregator nodes. In this work, a secure framework for environment monitoring using CPSs is presented. CPS controls and monitors the system using security algorithms, integrated with the internet and its users. This employs an outlier analysis technique to eliminate the false data injected by the compromised nodes including the aggregator nodes ensuring secure data aggregation. Further, this scheme provides confidential data transmission and effectively handles the other inside attacks. Simulation results show that about 95% of the injected false data are filtered from the network even if 25–30% of the sensor nodes are compromised.

6.1 INTRODUCTION

The exponential growth in the development and deployment of various types of cyber-physical systems (CPSs) has brought impact to various applications such as energy, transportation, environmental monitoring, military, healthcare, and manufacturing.[1] Wireless sensor networks (WSNs) are used in CPS for environment monitoring. WSNs comprises of large number of sensor nodes that are densely deployed in a geographical area where specific events have to be monitored or detected. These sensor nodes are low power devices with limitations on computing capability and storage capacity. These sensor nodes sense the occurrence of events and transmit the sensed data to the base station. Certain events may be detected by multiple sensor nodes which lead to data redundancy and the transmission of redundant data to the base station consumes bandwidth and power unnecessarily. The limited computing capacity of the sensor nodes require the reduction of data redundancy. Data aggregation is performed at the aggregator nodes to reduce data redundancy. The main task of an aggregator node is to gather data from the sensor nodes in the network and summarize the data based on a suitable function in order to reduce redundancy and transmit them to the base station.

Compromised sensor nodes inject false data into the network disrupts the data integrity in the process of aggregation in WSNs. This not only leads to spurious results at the base station, but also depletes battery power unnecessarily. In WSNs, the injected false data can be treated as outliers. Outliers refer to the readings or data that significantly deviate from the normal pattern.[2] There are many security schemes which address the issue of false data detection. The schemes presented in[3–6] address false data detection only during data forwarding. The schemes presented in[7–10] address secure data aggregation. Data Aggregation and Authentication (DAA) protocol[11] deals with false data detection during data aggregation as well as data forwarding. These schemes[3–11] do not address the other inside attacks like selective forwarding, Denial-of-Service (DoS) attacks etc. launched by WSNs.

Outlier detection is a technique used to detect the false data or erroneous data sensed by the faulty or compromised sensor nodes. The schemes presented in[12–15] deal with outlier detection in WSNs. In [16] some of the inside attacks in WSNs are addressed. In Ref. [17],[17] most of the inside attacks of WSNs are resolved by using a trusted cluster head/aggregator node. The aggregator node selection process used in,[17] cannot always guarantee the choice of trusted aggregator node. In practical situations, aggregator nodes can be malicious as well. The schemes[11,18] handle aggregator node compromises, but they do not address the issue of erroneous or false readings generated by faulty or compromised sensor nodes.

In this book chapter, an Outlier Analysis (OA) scheme is proposed to eliminate the false data injected by the compromised nodes including the aggregator nodes. This scheme effectively resolves the other inside attacks like selective forwarding, replay messages, Sybil attacks, DoS attacks, etc. and isolates those compromised nodes from the network. The detection and elimination of outliers in the sensor readings ensures data reliability and security. It also leads to the effective utilization of bandwidth and power. The isolation of such compromised nodes from the network offers better performance. Computation intensive cryptographic solutions will not suit WSNs due to its limited computing capacity. Hence, OA scheme is used to handle the inside attacks and eliminate the false data injected by the compromised nodes and to isolate the compromised nodes from the network.

The remainder of the book chapter is organized as follows: Section 6.2 deals with the research work related to false data detection, secure data aggregation and outlier detection; Section 6.3 describes the proposed OA scheme for eliminating the injected false data and isolating the compromised nodes from the network; Section 6.4 presents the performance evaluation of the proposed OA scheme; Section 6.5 gives the concluding observations and the future enhancements.

6.2 RELATED WORK

In this section, the research work is dealing with the detection of injected false data, secure data aggregation and OA in WSNs and CPSs are reviewed. In Ye et al.,[3] the statistical en-route detection and filtering scheme (SEF) facilitates the sensor nodes and the base station to detect false data with some reasonable probability. SEF scheme discards about 70% of the false data injected by the compromised sensor nodes within five hops and about 90% of the injected false data within ten hops. In,[4] false data injected by the compromised nodes are detected by the sensor nodes that collaborate to verify the integrity of the generated sensor data. In this scheme, sensor nodes are not allowed to carry out data aggregation while transmitting the data. The Commutative Cipher based en-route filtering scheme (CCEF)[5] discards false data without symmetric key sharing. In CCEF, the source node creates a secret association with the base station on a per-session basis and the intermediate forwarding nodes possess the witness key. The use of a commutative cipher enables the forwarding nodes to verify the authenticity of the data using the witness key. It does not require the original session key. In the dynamic en-route filtering scheme,[6] forwarding node can validate the authenticity of the sensor data only if it has an authentication key. The legitimacy report is endorsed by multiple sensor nodes using their unique authentication keys generated by one-way hash chains. The schemes[3–6] do not deal with the issue of fake message injection during data aggregation.

In Przydatek et al.,[7] proposed a scheme to verify the accuracy of the aggregated data at the base station using random sampling mechanisms. In,[8] the witness nodes aggregate data and compute MAC to verify the accuracy of the aggregators' data at the base station. The transmission of false data and MAC up to the base station depletes the scarce sensor network resources. Wu et al.[9] proposed a scheme in which topological constraints are introduced to build a Secure Aggregation Tree (SAT) that monitors data aggregator nodes. The sensor nodes use the cryptographic algorithms only when a malicious activity is detected. Yang et al.[10] proposed the Secure hop-by-hop Data Aggregation Protocol (SDAP) in which additional trust is placed on the nodes that are closer to the root compared to the other sensor nodes, during the aggregation process. In DAA scheme,[11] the false data injected in data aggregation and forwarding phase is detected.

Outlier detection and analysis is a critical research issue. In,[12] an aggregation tree is used to identify the faulty sensor nodes by setting the maximum and minimum values of the sensed attribute and their corresponding locations. Any sensor node that sends a data value outside this range can be identified as a faulty sensor node. In,[13] abnormal behavior of the sensor nodes is detected by computing a running average and comparing it with a threshold that is adjusted by a false alarm rate. This helps in defending the spoofing and sink-hole attacks. An in-network outlier cleaning approach proposed in[14] assures that outliers can be either corrected or removed from further transmission. Hawkins applied the naive Bayesian classification technique to detect outliers' misbehavior in WSNs.[15] This scheme detects outliers with high accuracy using the spatiotemporal correlations. Liu et al. proposed an insider attacker detection scheme (IADS) in WSNs.[16] This scheme is localized and suitable for large-scale WSNs. It exploits the spatial correlation of the sensor nodes in close proximity and achieves reasonable detection accuracy and low false alarm rate. But this scheme does not combat some of the inside attacks, such as data tampering or alteration, generating fake or false messages or eavesdropping of messages. The above inside attack issues are addressed in the Outlier Detection and Countermeasures Scheme (ODCS).[17] But ODCS does not deal with the compromises of the aggregator node or the cluster head; it assumes that aggregator nodes/cluster heads are trustworthy. But this assumption is a serious drawback of ODCS as the aggregator nodes can also be compromised which leads to spurious results at the base station. The schemes presented in[11,18] provide security against aggregator node compromises. DAA[11] combats the false data injected by multiple compromised sensor nodes including the aggregator nodes. But it does not consider other inside attacks. In,[18] three schemes are proposed to provide security in WSNs against aggregator compromises using multipath routing. But these schemes,[11,18] do not address the issue of the outliers in sensor readings generated by the compromised or faulty sensor nodes in CPSs. An outlier detection method using the K-means algorithm and big data processing is employed in.[20] Techniques using big data analytics extract meaningful results and construct intelligent models with help of machine learning tools.[20]

6.3 PROPOSED WORK

In this section, an OA scheme is proposed to remove the injected false data and resolve the other inside attacks launched by the compromised sensor nodes including the compromised aggregator nodes. The aggregator nodes are observed by "x" observer nodes to withstand aggregator node compromises. The OA scheme ensures the accuracy of the aggregated data. It guarantees the transmission of the correct data to the base station. Moreover, it effectively isolates those compromised nodes from the network. The performance of this scheme is evaluated and it is observed that this scheme achieves better false data filtering and low false alarm rate.

6.3.1 SYSTEM MODEL

The sensor nodes are deployed in the target field for environment monitoring as shown in Figure 6.1. The sensor nodes are grouped into clusters and an aggregator node is selected for each cluster using SANE Protocol.[21] It is a random aggregator node election scheme which resolves the adversarial attacks. To avoid battery drain, the aggregator node selection is randomized among the sensor nodes. Each sensor node in the cluster senses the information and transmits them to the corresponding aggregator node at regular intervals. The nodes in close proximity sense the same event due to the spatiotemporal correlation and transmit the data to the aggregator node. At any instance of time, a specific event can be sensed by "x" nodes in a cluster. Each node endorses the sensed data and sends them to the aggregator node of the cluster. The aggregator node summarizes the data received from all the sensor nodes in the cluster in order to reduce redundancy. In this system model, the sensor nodes other than the aggregator nodes are classified into three types: Neighboring nodes, Observer nodes, and Forwarding nodes. The compromised nodes are distributed uniformly in the network. Table 6.1 presents the summary of notations used in this book chapter.

The nodes closer to the aggregator node within the ith cluster are termed as the neighboring nodes and represented as N_i. The subset of nodes chosen from the neighboring nodes is termed as observer nodes of the ith cluster and is represented as O_i. The nodes that lie between any two aggregators are termed as forwarding nodes of the ith cluster and are represented as F_i. The aggregator node, the observer nodes, and forwarding nodes eliminate the false data injected into the network by checking the data packet. In this scenario, "x" observer nodes are selected from the neighboring nodes of the aggregator node in the cluster. Similarly "x" forwarding nodes are selected from those nodes which lie in the path between two aggregator nodes.[11]

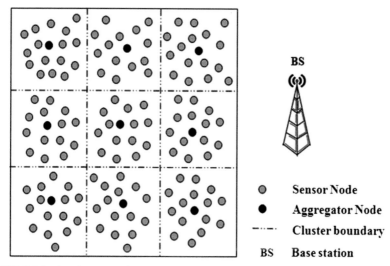

FIGURE 6.1 Deployment of sensor nodes in the network.

TABLE 6.1 Summary of Notations.

Notation	Meaning
ID	Identifier of a sensor node in a cluster
D_i	Data sensed by the ith sensor node in the cluster
K_{UG}	Group Key shared between the sensor node and the aggregator in the cluster
D_{agg}	Aggregated data
A_i	Aggregator node of the ith cluster.
N_i	Neighboring nodes of the aggregator node in the ith cluster
O_i	Observer nodes in the ith cluster
F_i	Forwarding nodes in the ith cluster
K_{ij}	Key shared between two aggregators A_i and A_j
$E_{Kij}(Dagg)$	Encryption of aggregated data Dagg with key K_{ij}
$MAC_{Kij}(Dagg)$	Message authentication code for the aggregated data Dagg with key K_{ij}

6.3.2 ATTACK MODEL

Compromised nodes in WSNs can launch various attacks. Typically, there are two types of attacks: (1) Exceptional message attack and (2) Abnormal behavior attack.[17] The former, attacks the network by tampering the message content or generating fake messages, such as false data injection. The latter, mainly attacks the network via dropping/forwarding messages to adversary or broadcasting redundant messages to waste energy and cause communication traffic, such as sink-hole, hello flood, DoS, and selective forwarding attacks.

6.3.3 KEY ESTABLISHMENT

In this scheme, it is necessary to establish the following keys: (1) Pair-wise key establishment between any two sensor nodes and (2) Group Key establishment between the sensor node and the aggregator node in a cluster.

6.3.3.1 PAIR-WISE KEY ESTABLISHMENT

Pairing of sensor nodes must be established among any two sensor nodes in the network[11] for pair-wise key establishment. The node pairs are formed between: (1) Aggregator of the ith and i + 1th cluster, (2) Observer node of the ith cluster and neighboring nodes of the i + 1th cluster and (3) Observer node of the ith cluster and forwarding node of the ith cluster.

There is one pair between the aggregator nodes of any two clusters. There are "x" pairs between observer nodes of the ith cluster and neighboring nodes of the i + 1th cluster. Similarly there are "x" pairs between the observer nodes of the ith cluster and the forwarding nodes of the ith cluster. The number of sensor node pairs created between any two clusters in the network is 2x + 1. This pairing of sensor nodes aids in handling x − 1 node compromises including the aggregator nodes by authenticating the data packet. If the data authentication fails even in any one node pair, then the data are removed considering it as false data. The packet format is shown in Figure 6.2.

Destination Address	Timestamp	Length	Source Address	$E_{KIJ}(Dagg)$	$MAC_{KIJ}(Dagg)$	$MAC(E_{KIJ}(Dagg))$

FIGURE 6.2 Packet format.

For pair-wise key establishment, random key distribution is applied. The existing random key distribution protocols presented in[22,23] allow key establishment using multi-hop communication. By using random key distribution protocols, the observer nodes can ensure the identity of their pair-mates, thereby preventing Sybil attacks where a compromised node fakes multiple identities to establish pair relations with more than one observer node. A random key distribution protocol is used against Sybil attacks if each node ID is associated with the keys assigned to that node and a key validation is performed before establishing pair-wise keys.[24]

6.3.3.2 GROUP KEY ESTABLISHMENT

The sensor nodes and the aggregator node of the cluster establish the cluster key K_{UG} using an existing group key establishment scheme.[25] The group key is used for selecting the observer nodes of the aggregator node, and protecting data confidentiality while data are transmitted among data aggregator and its observer nodes for data verification and aggregation.

6.3.4 OUTLIER ANALYSIS SCHEME

The sensor nodes deployed in the field are grouped into clusters. Each cluster has an aggregator node which summarizes the data and reduces data redundancy and forward the aggregated data to the base station through the subsequent aggregator nodes. The accuracy of the aggregated data depends on the legitimacy of the sensor nodes. Every sensor node in the cluster shares a unique key K_{UG} with the aggregator node. At regular time intervals, each sensor node in the cluster senses the data D_i from the target field. Any event occurring in the field is sensed and endorsed by "x" sensor nodes to avoid DoS attacks. Each data packet is time stamped to prevent replay attacks. Each sensor encrypts the node ID and the data with the key K_{UG} shared with the aggregator node in the following format $E_{KUG}(ID, D_i, t)$. Then this encrypted data are transmitted to the aggregator node of its cluster. The aggregator node decrypts the encrypted data sent by all the sensor nodes using the key K_{UG}. The aggregator node then computes the mean μ, median m_1-the central value, mode m_2-the most occurring value and the standard deviation s of the data transmitted by all sensor nodes. The formula for computing mean is given in eq 6.1. The median is calculated by computing the central value of the data values from the sensor nodes and the mode is the most frequently occurring value from the given sensor data values.

$$\text{Mean } \mu = \sum_{i=1}^{n} \frac{Di}{n} \tag{6.1}$$

The standard deviation s is calculated by the formula given in eq 6.2.

$$\text{Standard Deviation } s = \sqrt{\sum_{i=1}^{n} \frac{(Di-\mu)^2}{(n-1)}} \tag{6.2}$$

Based on these values, the central value x is calculated as given in eq 6.3.

$$x = \frac{\mu + m1 + m2}{3} \tag{6.3}$$

The permissible values of sensor readings are calculated as given in eq 6.4.

$$\text{Permissible range} = x \pm s \tag{6.4}$$

The values not within the permissible range are considered as false data and they are eliminated by the aggregator node. Then the aggregator node records the count of false data occurrence along with its node ID. If the false data occurrence count of a sensor node is greater than the threshold value then that sensor node is identified as compromised node. Then this compromised node is isolated and further data originating from this node is discarded at the aggregator node. If the sensor node data are within the permissible limit computed by the aggregator node then it is checked for redundancy, that is, multiple sensor nodes may detect the same event at the same time and hence can send the same data value. This data redundancy has to be eliminated. The aggregator node checks whether the sensor node value is the redundant entry of a specific event at the current instance of time. If it is a redundant entry then it is eliminated otherwise it is forwarded to the next aggregator node through the forwarding nodes. Figure 6.3 shows the algorithm for the OA scheme carried out at the aggregator node and Figure 6.4 gives the algorithm for handling aggregator node compromises. The frequency of sensing and transmitting data by the sensor nodes is monitored by the sensor node pairs. By exploiting the spatiotemporal correlation, the nodes in close proximity must have similar frequencies of transmitting data. A sensor node that transmits data or messages at a lesser or higher frequency than the specified limit, is identified as a compromised node for launching hello flood attacks, selective forwarding, sink hole, and wormhole attacks. Such nodes are isolated from the network by revoking the keys and discarding the data transmitted by them.

Algorithm: Outlier analysis scheme

Input: Readings collected at the aggregator node from the n sensor nodes D_i, where i = 1 to n.

Output: False data elimination and compromised node isolation.

1.   Each sensor node encrypts the sensed data D_i with key K_{UG} as $E_{KUG}(ID, D_i, t)$ and sends it to the aggregator node

2.   Aggregator node decrypts the sensor data from all the sensor nodes associated with it

3.   Compute mean μ, median m_1, mode m_2, and the standard deviation s of the sensor readings

4.   Compute $x = (\mu + m_1 + m_2)/3$

5.   Calculate the permissible values of sensor readings as $x \pm s$

6.   **if** the value from the sensor node is within the permissible range **then**

7.   **if** the sensor node value is redundant entry of a specific event **then**

8.   Eliminate the redundant value

9.   **else**

10.   Forward it to the next aggregator

11.   **end if**

12.   **else if** the value is not within $x \pm s$ **then**

13.   Eliminate the data value and increment the False Data Count of that sensor node ID by 1

14.   **if** the False Data Count of that sensor node ID > Threshold limit **then**

15.   Isolate the node from the network by discarding further data from that node

16.   **end if**

17.   **end if**

FIGURE 6.3   Algorithm for outlier analysis.

Algorithm: Handling aggregator node compromises

Input: A set of aggregators A_{i-1}, A_i and A_{i+1}, 'n' neighbouring nodes, 'x' observer nodes and 'x' forwarding nodes of each aggregator

Output: False data injected by x-1 compromised sensor nodes is eliminated.

1. Initialize i=2
2. **while** $(i < R)$ **do**
3. Encrypt data D using the key $K_{i-1, i}$ as $E_{Ki-1,i}(D)$
4. Transmit $E_{Ki-1,i}(D)$, MAC($E_{Ki-1,i}(D)$), MAC(D) to A_i from A_{i-1}
5. Decrypt $E_{Ki-1,i}(D)$ to get data D
6. Encrypt D as $E_{KUG}(D)$ using the group key K_{UG}
7. Transmit $E_{KUG}(D)$ and MAC(D)
8. **if** authentication is successful at O_{i-1}-N_i pair **then**
9.   forward the data packet
10. **else**
11.   discard the data packet
12. **end if**
13. Encrypt the data D as $E_{Ki,i+1}(D)$ using key $K_{i,i+1}$
14. Observer node computes MAC($E_{Ki,i+1}(D)$) and MAC(D) and sent to A_i
15. Compute MAC($E_{Ki,i+1}(D)$) and MAC(D) for verification at A_i.
16. **if** authentication is successful at A_i **then**
17. Create a packet with $E_{Ki,i+1}(D)$, MAC($E_{Ki,i+1}(D)$) and MAC(D) and is sent to A_{i+1}
18. **else**
19.   Discard the data
20. **end if**
21. **if** sensor node is not within O_i-F_i pair **then**
22.   Forward data packet to the next forwarding node
23. **else**
24.   MAC($E_{Ki,i+1}(D)$) is checked at F_i
25. **end if**
26. **if** authentication is successful at O_i-F_i pair **then**
27.   Forward the data packet to A_{i+1}
28. **else**
29.   Discard data packet and inform A_i
30. **end if**
31. Increment i by 1
32. **end while**

FIGURE 6.4 Algorithm for handling aggregator node compromises.

6.4 PERFORMANCE EVALUATION

This work was carried out by simulating a WSN for temperature monitoring system using C++ and MATLAB. The simulation parameters are listed in Table 6.2. The performance of the proposed OA scheme is evaluated by comparing it with ODCS.[17] The performance of the OA scheme is evaluated based on security, computational overhead, and communication overhead.

TABLE 6.2 Simulation Parameters.

Simulation parameters	Values
Simulation area	(0,0) to (300,300)
Number of sensor nodes	300
Number of base station	1
Base station	(350,150)
Initial energy	2 Joules
Packet size	512 bytes
Cluster radius	40 m

6.4.1 SECURITY ANALYSIS

The metrics used to evaluate the security of the OA scheme are: (1) Probability of false data detection and (2) False alarm rate. The filtering capacity is dependent on the probability of false data detection. The false alarm rate refers to the probability of discarding the legitimate data by the sensor nodes as false data.

The OA scheme is effectively removed the false data by verifying the MAC of the data packet at the sensor node pairs due to the following reasons: the neighboring nodes of A_i authenticate the data broadcast by A_i and, therefore, A_i cannot inject any false data. The observer nodes of A_i compute MAC and any false data injected by A_i will be eliminated by the neighboring nodes of A_{i+1}. The "x" O_i-F_i pairs also authenticate the MAC of the data packet and eliminate false data. The value of "x" is chosen based on the density of node deployment and the security requirement. This indicates that the scheme can effectively guard against x-1 sensor node compromises. Apart from false data elimination, this scheme can protect against replay attacks and DoS attacks. The sequence number along with the timestamp/nonce in the data packet helps in eliminating the replayed messages. The compromised nodes can encrypt the data packets with a legal key and launch DoS attacks. In this scheme, there are atleast "x" sensor nodes in a cluster that sense a single event and each sensor node endorse the sensed data and transmit them to the aggregator node. Therefore, the DoS attack instigated by x-1 compromised nodes can be defended successfully. Anscombe's quartet[26] used mean and variance. But the proposed approach makes use of mean, median, and mode along with standard deviation which helps in improving the effectiveness of the algorithm.

An efficient outlier detection technique must have very high detection rate and very low false alarm rate. A Receiver Operating Characteristic (ROC) curve is used to represent the tradeoff between detection rate and the false alarm rate.[27] False alarm rate is otherwise called as False Positive Rate. This refers to an alarm or a notification when there is no malicious activity like false data injection. If the area under the ROC curve is large, then the detection technique offers better performance.[27,28]

The ROC curve in Figure 6.5 represents the tradeoff between detection rate and the false alarm rate for the ODCS and the proposed OA scheme. The graph also shows the ideal curve where there is no false alarm notification. The ideal curve is practically impossible to achieve in real-time scenario. The area under the ROC curve represents its performance, that is, larger the area higher the performance. The proportion of false alarm notification is very less in the proposed OA scheme compared to ODCS. The detection and elimination of false data improve the accuracy of the aggregated data transmitted to the base station.

The probability of detecting false data injected in the network is evaluated by taking into consideration the impact of number of nodes in the cluster and the number of nodes in the network. The OA scheme detects and filters about 95% the injected false data when compared to the ODCS which filters about 90% of false data[17] in the presence of 25% of the compromised nodes. This comparison is represented in the graph shown in Figure 6.6(a) and (b). Further, the proposed OA scheme filters 95% of the false data at lesser hops than ODCS. The filtering capacity, that is, the probability of detecting the false data is analyzed with respect to the scalability of nodes. The increase in the number of nodes in the cluster or the network does not affect the filtering capacity.

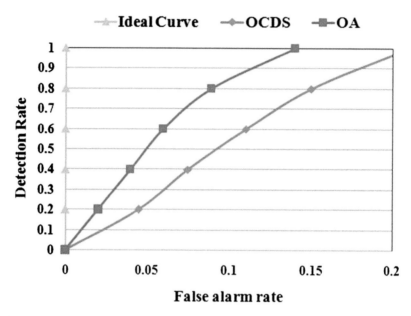

FIGURE 6.5 ROC curve.

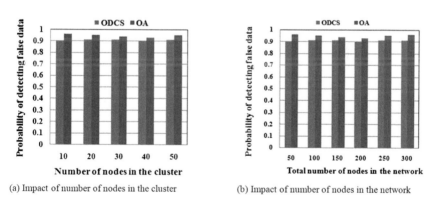

(a) Impact of number of nodes in the cluster

(b) Impact of number of nodes in the network

FIGURE 6.6 Probability of false data detection.

The probability of the injected false data which is not detected by the ODCS and the proposed OA scheme is shown in Figure 6.7(a) and (b). This is again plotted against the number of nodes in the cluster and the number of nodes in the network. The probability of not detecting the false data for the proposed OA scheme is only about 0.05. For the ODCS scheme, this probability is 0.1.

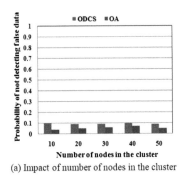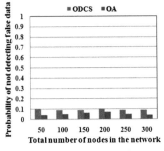

(a) Impact of number of nodes in the cluster　　　(b) Impact of number of nodes in the network

FIGURE 6.7　Probability of false data not detected.

False alarm refers to the detection and filtering of legitimate data by the sensor nodes as false data. From Figure 6.8, the impact of the number of nodes in the network on the probability of the false alarm rate for both the schemes is shown. The false alarm rate for ODCS is around 0.1 and for OA scheme is 0.05. This implies that OA scheme has lower false alarm rate than ODCS.

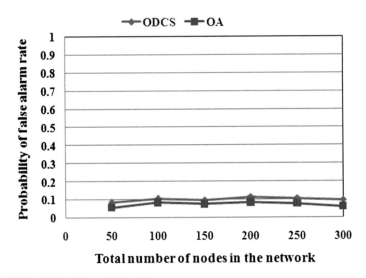

FIGURE 6.8　Probability of false alarm rate.

The impact of the ratio of false-to-legitimate data on false data detection probability for both ODCS and OA schemes are compared. The ratio of false-to-legitimate data represents the false data that is injected in the network along with the legitimate data. It is seen from Figure 6.9 that the probability of false data detection is above 90% till the ratio of false-to-legitimate data is 0.43. After this, the probability of detecting the false data decreases as the ratio of false-to-legitimate data increases. This implies the increase in the false data injection affects the filtering capacity. The value 1 indicates that the number of false data is equal to the number of legitimate data. The value 0.43 represents that 30% of the data is false data which implies there are 30% compromised nodes.

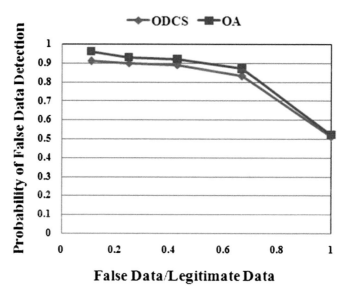

FIGURE 6.9   Impact of the false-to-legitimate data ratio on false data detection probability.

The proposed OA scheme ensures the accuracy of the aggregated data. It guarantees the transmission of the legitimate data to the base station. Moreover, it effectively isolates those compromised nodes from the network. The performance of this scheme is evaluated and it is observed that this scheme achieves better false data filtering and low false alarm rate. Further, the proposed OA scheme handles aggregator node compromises and it is immune to most of the inside attacks. Table 6.3 presents the comparison of the two schemes – ODCS and OA scheme against the various attacks.

TABLE 6.3   Immunity against Attacks.

Attack type	ODCS	OA
Aggregator node compromise	×	√
False data injection	√	√
Selective forwarding	√	√
Sinkhole attack	√	√
Sybil attack	√	√
Worm-hole attack	√	√
Hello flood	√	√

6.4.2   COMPUTATIONAL OVERHEAD AND COMMUNICATION OVERHEAD

The computational overhead is evaluated for both the schemes by the total number of MAC computations involved in the scheme. The impact of the number of nodes in the network on the MAC computations is shown in Figure 6.10. It is observed that the computational overhead for the OA scheme is higher than that of the ODCS. This is because of the additional MAC computations involved in handling aggregator node compromises.

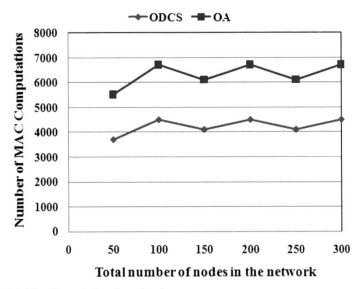

FIGURE 6.10   Computational overhead.

The communication overhead is evaluated for both the schemes by the total data transmission in the network. The impact of the number of nodes in the network on the total data transmission is represented in Figure 6.11. It is observed that the communication overhead for the OA scheme is higher than that of the ODCS. This is because of the additional messages involved in handling aggregator node compromises. There is about 2–3% increase in communication overhead in OA scheme compared to ODCS.

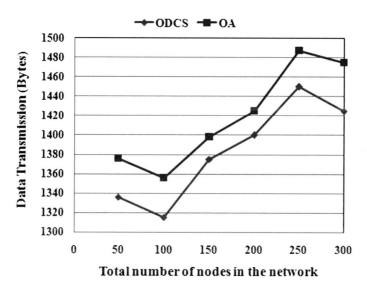

FIGURE 6.11 Communication overhead.

6.5 CONCLUSION

This main focus of this book chapter is to ensure security in the process of data aggregation and forwarding for environmental monitoring using CPSs. This is achieved by eliminating the injected false data and resolving the other inside attacks like selective forwarding, Sybil attacks, DoS attacks, etc. launched by the compromised sensor nodes including the compromised aggregator nodes. This scheme effectively handles x-1 node compromises and isolates those compromised sensor nodes from the network. The false data injected by the compromised sensor nodes are detected and eliminated through OA and thereby ensuring data integrity.

This scheme eliminates about 95% of the injected false data from the network. Simulations also indicate that the scheme works effectively even if 25–30% of the sensor nodes are compromised. The additional computational and communication overhead involved in OA scheme due to the handling of aggregator node compromises is inevitable for providing security. As future work, it is planned to reduce the computational and communication overhead involved in this scheme.

KEYWORDS

- **cyber physical systems**
- **wireless sensor networks**
- **false data injection**
- **inside attacks**
- **secure data aggregation**
- **compromised nodes**
- **outlier analysis**

REFERENCES

1. Humayed, A.; Lin, J.; Li, F.; Luo, B. Cyber-physical Systems Security—A Survey. *IEEE Internet Things J.* **2017,** *4* (6), 1802–1831.
2. Zhang, Y.; Meratnia, N.; Havinga, P. Outlier Detection Techniques for Wireless Sensor Networks: A Survey. *IEEE Commun. Surveys Tutorials* **2010,** *12* (2), 159–170.
3. Ye, F.; Luo, H.; Lu, S.; Zhang, L. Statistical En-route Detection and Filtering of Injected False Data in Sensor Networks. *Proc. IEEE INFOCOM* **2004,** *4*, 2446–2457.
4. Zhu, S.; Setia, S.; Jajodia, S.; Ning, P. Interleaved Hop-by-hop Authentication against False Data Injection Attacks in Sensor Networks. *ACM Trans. Sens. Netw.* **2007,** *3* (3).
5. Yang, H.; Lu, S. Commutative Cipher Based En-route Filtering in Wireless Sensor Networks. *Proc. IEEE VTC* **2004,** *2*, 1223–1227.
6. Yu, Z.; Guan, Y. A Dynamic En-route Scheme for Filtering False Data in Wireless Sensor Networks. *Proc. IEEE INFOCOM* **2006,** 1–12.
7. Przydatek, B.; Song, D.; Perrig, A. SIA: Secure Information Aggregation in Sensor Networks. *Proc. SenSys'03* **2003,** 255–265.
8. Du, W.; Deng, J.; Han, Y. S.; Varshney, P. K. A Witness-based Approach for Data Fusion Assurance in Wireless Sensor Networks. *Proc. GLOBECOM'03* **2003,** 1435–1439.

9. Wu, K.; Dreef, D.; Sun, B.; Xiao, Y. Secure Data Aggregation Without Persistent Cryptographic Operations in Wireless Sensor Networks. *Ad Hoc Netw.* **2007,** *5* (1), 100–111.

10. Yang, Y.; Wang, X.; Zhu, S.; Cao, G. SDAP: A Secure Hop-by-hop Data Aggregation Protocol for Sensor Networks. *Proc. ACM MOBIHOC'06* **2006.**

11. Ozdemir, S.; Cam, H. Integration of False Data Detection with Data Aggregation and Confidential Transmission in Wireless Sensor Networks. *IEEE/ACM Trans. Netw.* **2010,** *18* (3), 736–739.

12. Banerjeea, T.; Xiea, B.; Agarwal, D. P. Fault Tolerant Multiple Event Detection in a Wireless Sensor Network. *J. Parall. Dist. Comput.* **2008,** *68* (9), 1222–1234.

13. Li, D.; Wong, K. D.; Hu, Y. H.; Sayeed, A. M. Detection, Classification, and Tracking of Targets *IEEE Signal Process. Mag.* **2002,** *19,* 17–29.

14. Zhuang, Y.; Chen, L. In-network Outlier Cleaning for Data Collection in Sensor Networks. In *Proc. 1st Int. VLDB Workshop Clean Databases CleanDB'06*; Seoul, Korea, September 2006; pp 41–48.

15. Hawkins, D. *Identification of Outliers*; Chapman and Hall, 1980.

16. Liu, F.; Cheng, X. Z.; Chen, D. C. Insider Attacker Detection in Wireless Sensor Networks. In *INFOCOM 2007 26th IEEE Int. Conf. Comput. Commun.*; Anchorage, Alaska, USA, May 2007; pp 1937–1945.

17. Zhang, Y.-Y.; Chao, H.-C.; Chen, M.; Shu, L.; Park, C.-H.; Park, M.-S. Outlier Detection and Countermeasure for Hierarchical Wireless Sensor Networks. Info. Security, IET **2010,** 4 (4)

18. Claveirole, T.; Dias de Amorim, M.; Abdalla, M.; Viniotis, Y. Securing Wireless Sensor Networks against Aggregator Compromises. *IEEE Commun. Mag.* **2008,** 134–141.

19. Souza, A. M.; Amazonas, J. R. An Outlier Detect Algorithm Using Big Data Processing and Internet of Things Architecture. *Proc. Comput. Sci.* **2015,** *52,* 1010–1015.

20. Parwez, M. S.; Rawat, D.; Garuba, M. Big Data Analytics for User Activity Analysis and User Anomaly Detection in Mobile Wireless Network. *IEEE Trans. Ind. Info.* **2017,** *13* (4), 2058–2065.

21. Sirivianos, M.; Westhoff, D.; Armknecht, F.; Girao, J. Non-manipulable Aggregator Node Election Protocols for Wireless Sensor Networks. *Proc. Int. Symp. Model. Optim. Mob. Ad Hoc Wireless Netw. (WiOpt)* **2007.**

22. Du, W.; Deng, J.; Han, Y. S.; Varshney, P. K. A Pairwise Key Predistribution Scheme for Wireless Sensor Networks. *Proc. 10th ACM CCS* **2003,** 42–51.

23. Liu, D.; Ning, P.; Li, R. Establishing Pairwise Keys in Distributed Sensor Networks. *ACM Trans. Inf. Syst. Security* **2005,** *8* (1), 41–77.

24. Newsome, J.; Shi, E.; Song, D.; Perrig, A. The Sybil Attack in Sensor Networks: Analysis and Defenses. *Proc. 3rd IEEE/ACMIPSN* **2004,** 259–268.

25. Blundo, C.; Santis, A.; Herzberg, A.; Kutten, S.; Vaccaro, U.; Yung, M. Perfectly-secure Key Distribution for Dynamic Conferences. *Proc. Crypto* **1992,** 471–486.

26. Anscombe, F. J. Graphs in Statistical Analysis. *Am. Stat.* **1973,** *27* (1), 17–21.

27. Lazarevic, A.; Ozgur, L.; Ertoz, J.; Srivastava, A. K. A Comparative Study of Anomaly Detection Schemes in Network Intrusion Detection. *SIAM Conf. Data Min.* **2003.**

28. Altay Guvenir, H.; Kurtcephe, M. Ranking Instances by Maximizing the Area under ROC Curve. *IEEE Trans. Knowledge Data Eng.* **2013,** 25 (10).

CHAPTER 7

RTLS: An Introduction

M. SENTHAMIL SELVI[1*], K. DEEPA[1], S. BALAMURUGAN[2],
S. JANSI RANI[1], and A. MOHAMED UVAZEAHAMED[3]

*1Sri Ramakrishna Engineering College, Coimbatore 600022,
Tamil Nadu, India*

*2Founder & Chairman–Albert Einstein Engineering and Research
Labs (AEER Labs); Vice Chairman–Renewable Energy Society of India
(RESI), India*

*3Department of Computer Science, Cihan University-Erbil,
Kurdistan Region, Iraq*

*Corresponding author. E-mail: senthamilselvi@srec.ac.in

ABSTRACT

A real-time location system (RTLS) enables you to find, track, measure, and analyze key data capable products are used in many sectors including Supply Chain Management (SCM), healthcare, military, retail, recreation, education and almost every business. RTLS, uses a tag, small wireless devices to locate and track people and resources. For example, you may attach a tag to any merchandise in the company so that you can locate it within the campus. RTLS first emerged in the 1990s used to distinguish and describe about the technology used for automatic identification of radio frequency identification (RFID) tags and provided the facility to locate a tagged asset on the computer screen. Earlier, these technologies were used by military and government agencies but it is very expensive for commercial applications.

7.1 REAL-TIME LOCATION SYSTEM

In today's market, "RTLS" is growing. Based on the survey taken and reported by global technology research organization and advisory firm Technavio, titled "Global Real Time Location Systems Market, for the year 2014–2020," the market growth will be 38% due to increasing acceptance and adoption of RTLS in healthcare.

Let's examine these terms in detail[1]

Let's explore the term RTLS,

- Real time: Location of person/assets is readily available or with short latency
- Location: A symbol/tone/other methods used to point the position of the person or asset, which is used to relate them to exact environment
- System: System with hardware and software technologies to gather, process, and deliver the information in an organized and structured manner
- RTLS is usually embedded with mobile phones or any navigational systems.

7.1.1 RTLS APPLICATIONS

RTLS applications include:

Fleet tracking: Enterprise uses the RTLS systems to track the driver of the vehicle, vehicle speed, location, optimize the routes to reach the destination, check the schedule jobs and aid navigation.

Navigation: RTLS system provides the assistance for navigation which gives the directions to locate Y destination from X source. Complex navigation services provided when we incorporating GPS, mobile technology, and mapping.

Human resource tracking: Locating the human resources be very difficult, if the organization is large this leads the poor communication and development of enterprise. Human resources in enterprise is tracked by GPS enabled devices (especially mobile device). RFID-enabled badges used to track on-site workers.

Inventory and asset tracking: Enterprises uses RFID tags for communicate wirelessly with RFID readers.

Network security: Used to restrict the access based on the user location. The physical border area is formed by using Wi-Fi protected access.

7.2   INDUSTRIAL APPLICATIONS OF RTLS[2]

Based on three major applications such as

- ➢ **Healthcare:**
  - RTLS implementation gives a lot of stress relief for hospital management, because it is used to track the staff, patients, and equipments. In addition to that the system help to improve the inpatient and outpatient management, intensive care unit (ICU), emergency department, and post-anesthetic care unit, etc.
    - Medical Equipment Tracking: During an emergency situation, the tracking of the medical equipment is difficult; this is solved by RTLS system, and it improves the response for that situation and also improves the quality of the hospital in terms of care of the patients.
    - Patient Protection: Used to track elderly patients by hospital staffs who wander off, also used child care.
    - Locate Medical Staff: The infrastructure of the hospital is very much improved and area is also large. The more number of the staff is working in shift basis; the more difficult it is to locate the staff during needed time. RTLS provides solution, tags are fitted with buzzers and LED for communication during the emergency situation.
- ➢ **Manufacturing:**
  - Machines, manpower, and materials are the pillar of the manufacturing sector. The above-mentioned things are important for the investor to improve the production and customer service; this can be done by track on real-time basis. RTLS is adapted to the manufacturing sector based on the demand on the reduction of operational costs.
    - Process Tracking: Key performance indicator for the manufacturing sector is very important it can be monitored and eliminated the deficiency immediately by RTLS. It also

gives an efficient and effective environment for production in terms of assembly-line process.

- Movement Tracking: The equipment and workers of the industry can be tracked during the production and used to place and replace the equipment/workers to improve the productivity in the respective sector.
- Employee Safety: During production, human works close to the machines they even do not know about the safe distance while in working, this situation can be avoided by RTLS safety measures by giving the alter message to the workers.

➢ **Retail:**
- Usage of RTLS is not only limited to the above sectors, but also used in retail industries. The system used in their store to increase the business based on the customer behavior in store experience. Also, provides customized customer service in both online/offline stores.
  - Customer Analytics: The traditional way to analyze the customer pattern is by asking questionnaire. This is very time consuming and lot of human efforts required. RTLS solution provides information about the customer by analyzing the customer such as visits to the store, location visit in the store, stay time without affecting the customer privacy.
  - Store layout optimization and catalog designing: In current scenario, customer satisfaction is very important. RTLS will improve the store layout, retail exposure, and dwell times of the customer with customized and effective catalog design.
  - Targeted Ads: Marketing of the product is very essential in today's competitive world. RTLS system provides location-based advertise and immediacy campaigns, which help to deliver the contextual content and make the customer to visit and shop in the store.

7.3   REASONS- NEED OF RTLS[3]

RTLS used to control the assets in large area.

1. Easy identification of misplaced asset:
   The exact location of the resources easily viewed by RTLS, will eliminate the time-consuming searching process and audit.

It does not need any requirement, location data integrate into the existing ERP system of the enterprise; it allows utilizing the more value from their IT infrastructure with little effort.

2. Reduce equipment movement:

Based on the requirements of the enterprise, equipment relocated from place to place. It requires a lot of human and machine effort and leads mishandling of the equipment. Real time minimizes the relocation. Instead of moving, equipment placed closer to the point of use, which turns improvement in production rate.

3. Simple audit and lost property recovery:

Audits of asset and workers are simple with real-time, that is, working of the workers and equipment lost can be monitored and detected quickly.

4. Improve inventory and warehouse management:

RTLS automates the inventory control process like choose and put-away location, shipping verification, data entry process, whereas provides on-time and faster delivery of the products and increase the customer satisfaction.

5. Inbound and outbound logistics:

Tracking the moving of the asset in and around the industrial complex is very multifaceted, it is eliminated by RTLS. The result is higher efficiency of the storage areas and equipments and speed up the process and execution.

6. Transport management:

The movement of the transports like car, truck, trains, and vessel movement with tool sequence, constraint validation and shipment validation be planned and controlled. It increases the responsiveness throughout the port and terminal operations.

7. Capture disparate data:

ERP system in the enterprise integrated with middleware that reads auto-ID technologies like RTLS, RFID, and barcode and can quickly capture, filter, and disparate data. Real-time information captured and processed to make positive business decisions.

8. Improve cycle time:

   Used to reduce inventory cycle time to minutes against hours, days, or weeks by track items and personnel movement in zone and processes. Earlier these can be done manually by managers and process engineers.

9. Create ROI opportunities:

   Track asset movement and personnel speed up the tasks and transactions, reduces order-to-cash cycle, good forecasting of business, improve asset values, and investor satisfaction.

10. Improve safety and security:

    The major problems in traditional tracking technologies in RFID and GPS are interference, short range, and slow data rates. It causes safety and security issues by slow transmit rate while walking in large area also miss key information. RTLS enables alarm for locating the personal and how long they were present in the location. Simply, RTLS for "Need to Know" system.

7.4   USECASES OF RTLS[4]

Real-Time Location describes different things to different people depending on the proposed application. We will look at a few of the most common types here before considering how they might be implemented.

7.4.1   PROXIMITY

Proximity is the nearness or closeness of the object. Nodes used to found how far apart the object from each other and accordingly taken the actions. It reduces time to determine the distance between two items.

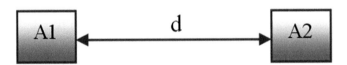

FIGURE 7.1   Proximity.

Example of "Proximity" application:

The nearness of patient or new born babies to an unlocked door through which they are not authorized to pass.

7.4.2   ABSOLUTE LOCATION USING FIXED INFRASTRUCTURE

Tag fitted to the objects/persons, location establishment of the tagged persons/objects based on the fixed anchors in known locations. This is traditional RTLS system.

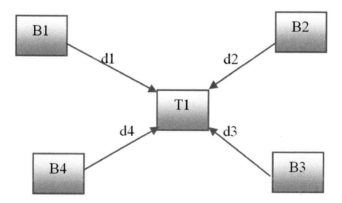

FIGURE 7.2   RTLS with fixed infrastructure.

Example includes:

The tracking and location identification of equipments and patients in healthcare:

- Efficient and effective patient care
- On-time staff during needed time
- Improved operating cost reduction

The tracking and location of items in warehousing and inventory help to :

- Reduced down time in production
- Improved Customer stratification
- Reduction – operational cost and man work

The tracking and monitoring of animals in the farm leading to:

- Increase animal yield and improve its health
- Improvement in terms of assistance for animals to reach owners

Manufacturing:

- The tracking of goods, work progress and finished goods and products
- Assembly of the components
- To monitor the movements of the tools to check whether works carried out in the correct sequences

7.4.3   RELATIVE LOCATION AMONG A GROUP OF NODES

There is no fixed infrastructure, so nodes must establish their location relative to other nodes in the network.

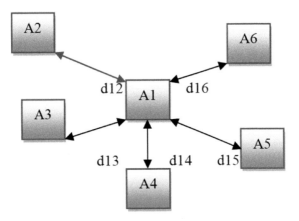

FIGURE 7.3   Relocation among group of nodes.

7.5   UNDERSTAND RTLS IN HEALTHCARE[5]

The benefit of the RTLS in healthcare starts from simple locating aspect to patient and management satisfaction. Different RTLS technologies are used by different vendors to take over and carry on in the market.

How to choose right RTLS for their own clinical environment? Here are some of the considerations that should be taken into account for RTLS:

1. The accuracy of the RTLS technology
2. Software
3. Support
4. Existing customer base

7.5.1 RTLS TECHNOLOGY REVIEW

Overview of types of RTLS technologies available in healthcare:

*   Infrared (IR),
*   RFID,
*   Ultrasound,
*   Ultra-wide band (UWB),
*   Wi-Fi, and
*   Zigbee

IR: Availability of RTLS for healthcare, IR is usually combined with RFID and deployed in two mechanisms. First, one-time cable installation in which a ceiling-mounted sensor is placed in each room or in each area that must be defined as a location (i.e., over a bed, in a storage closet, etc.). The badge emits IR signals and RFID signals, which are picked up by the sensor. The second IR/RFID system is a beacon IR system (called "reverse IR") where the badge serves as the "sensor" and receives IR signals from a ceiling-mounted monitor (sold as battery-operated).

Ultrasound: Sound waves, imperceptible to the human ear, are emitted from a tag and received by in-room, hardwired microphone detectors (sensors) tuned to detect the tag's ultrasonic frequency.

UWB: UWB utilizes short pulses of RF, which are emitted from a badge and received by a series of RF-like antennas and readers (sensors), which construct up the wireless mesh network.

Wi-Fi: Wi-Fi RTLS systems use an 802.11a/b/g/n wireless mesh network. Tags have IP addresses (or MAC addresses), which are routed to access points (sensors) located throughout the facility. Wi-Fi locating networks will require at least three times as many access points as the classic data communication network.

Zigbee: Zigbee, which also utilizes a wireless mesh network, though on the 802.15.4 layer which restricts the reach of the network communication area. Tags have MAC addresses that are transmitted to sensors and plugged into any standard AC outlet.

Each of these technologies will include at least basic software to translate the raw location data into a format understood by computer software The IR systems have need to read from a single point to define location. The others calculate location based on the received signal strength indicator (RSSI) or time difference of arrival (TDOA) algorithms.

Some of these RTLS vendors have software that displays an item's location or allows to search for an item, while others vendors will rely on third-party or partner vendors to interpret and display the data location obtained with the RTLS. Very few of these healthcare RTLS vendors offer RTLS technology and comprehensive software to support clinical RTLS applications. The issue of multiple vendors to support what is meant to be a consistent RTLS system is another area has to consider. What kind of support model is offered by vendor? If multiple vendors are involved, if something goes wrong?

Other questions to ask include:

- If the system requires recalibration, who is responsible (in terms of vendor)?
- If the system requires battery change, how often, how much, and who is responsible?
- Is the system self-aware? Can it locate network issues?
- Who provides the training of the system and how in-depth is it?

7.6   IMPLEMENTATION OF RTLS IN HEALTHCARE

This chapter provides an outline of RTLS components,[6] key healthcare requirements to consider and provides an overview of each of the main technology applications currently offered in the healthcare sector.

7.6.1   *RTLS COMPONENTS*

RTLS made up of various components. This section describes those components in detail.

7.6.1.1   ASSET TAGS

The most visible component is asset tag. A tag comes in various sizes and it can be customized for staff, patients, and also based on the equipments. RTLS tags are power-driven; therefore, battery life is a factor in operating costs. To identify the tag status and vary tag transmission intervals in order to reduce power consumption and extend battery life vendors make use of various schemes like motion detectors. The other features of the tags are switches, buttons, indicators, and tampering (unauthorized removal) indicators.

7.6.1.2   RECEIVERS

Receivers referred as sensor, detectors, or readers. To identify the tag location, tags must communicate with receivers. Receivers also powered like tags and positioning data delivered to positioning engine through enterprise LAN.

7.6.1.3   UNDERLYING TECHNOLOGY

The requirement of a number of receivers for the area with positioning accuracy can be fixed based on the technology shared between receivers and tags. "Bleed through" of tags between the area and accuracy of the tag location influenced by system design and technology. Some vendors employ more than one type of technology and receivers to improve location accuracy performance.

7.6.1.4   POSITIONING ENGINE

The position of the tags analyzed by the data generated by tags and receivers (receivers sent to the positioning engine). This is the software applications that can be packaged as an appliance or an application. Appliances combine the application software, already installed and configured on hardware. Appliances are usually installed in a data center or wiring closet, connected to the network, and managed remotely using a web browser. Applications are usually distributed on CD, and installed on general purpose server hardware provided by the RTLS vendor, the hospital or a third party. Software applications frequently require a few system management be done via a keyboard and display connected to the computer running the application.

7.6.1.5   TAG POSITIONING METHODS

Tag position is determined by two methods. First, zone-based system utilizes the receivers to record the tag passes the receiver, also this system records the direction of tag movement and its speed. Density of receivers influences the positioning accuracy of the zone-based system. Transmitted data from receivers use mathematical methods or formulas to identify the tag position. Second, the computational positioning system uses different methods like RSSI, TDOA, or RF fingerprinting. These two approaches for determining positioning, zone based and computational, are not mutually exclusive and some systems make use of both techniques.

7.6.1.6   POSITIONING ENGINE LATENCY AND SCALABILITY

Overall performance of the system and scalability based on the positioning engine and related software. Tag read rates and positioning engine performance influence the latency of the system. A tag read rate of more than several seconds result in positioning latency. The calculations to identify the location of the tag can vary from system to system and technology. Scalability refers how the implemented system "scaled up" or adopted for new changes and environment of the entire enterprise. The cost, scaling, and complexity also vary from system to system.

7.6.1.7   ALERTS, ALARMS, AND WORKFLOW MAPPING

7.6.1.7.1   RTLS System Track

RTLS software alerts and alarm produced in addition to determining and tracking the tags. These are generated from the information about the relationships between tagged objects, assets and their respective locations. Rules generated like an alert in response or by signaling equipment to improve the patient assistances or during surgical operations.

7.6.1.7.2 Third Party Application Integration

RTLS also provides the means to pass positioning data to third party applications. Simply, referred as application programming interfaces (APIs), a standard interface is provided by RTLS vendors so that specialized applications can utilize real-time positioning data. Examples of applications that can utilize an RTLS API include Emergency Department management systems, surgical management systems, and hospital-wide patient flow and bed management applications.

7.6.2 KEY HEALTHCARE REQUIREMENTS

In addition to the specific needs to which RTLS are applied, there are characteristics of the healthcare environment that impact RTLS evaluation and selection.

- Mobility: The delivery of healthcare is naturally mobile–few people or things maintain a static location. And those people and assets that do remain in a fixed location require significant effort to ensure optimal utilization.
- Change: Healthcare is always changing based on responses to new regulations or market requirements. In the USA (also in India), the hospital industry is undergoing a record building boom in new facilities, additions, and renovations. Because hospitals continually fine tune their physical plant, a RTLS must be able to accommodate this change.
- Physical environment: Environment of the hospital is a challenge because of diversity of hospital materials when compared with commercial office spaces. A hospital represents a very challenging environment mostly for RTLS. The variety of electromechanical equipment and large metal carts, fluids and other consumables also create challenges to RTLS. Implementation requires extensive renovations of the buildings and it needs more time.

7.6.3   RTLS COMPARISON CRITERIA

With the complex requirements and needs of healthcare environment, it is difficult to determine the characteristics that are important with different RTLS. The underlying technology used by individual RTLs is important also how a technology is implemented on design decisions and tradeoffs and the resulting impacts the comparison criteria below will indicate the best RTLS for a given situation.

7.6.3.1   ACCURACY

Knowing the location of people and asset is what positioning systems regards. Different positioning applications need different degrees of positioning resolution. Is the infusion unit fourth wing, or in patient room 402? Each of these degrees of positioning resolution may be fit, depending on operational goals. Before a selection of RTLS systems, the required degree of accuracy must be determined. Achieving a certain degree of positioning accuracy is the result of the capabilities of the positioning system as well as the design and implementation of the installed system.

7.6.3.2   RELIABILITY

The important characteristic of positioning accuracy is reliability; how often is a given tag position determined within the specified accuracy range? The design, implementation, and management of the RTLS can influence reliability. The system design must support performance requirements. The implementation phase of RTLS frequently requires an iterative fine tuning of receiver density and placement to ensure consistent and reliable performance. Knowing how a particular solution calculates asset location and any inherent latency with this process is also a key factor. Hospitals are challenging surroundings for technologies using RF. The wide range of hospital building materials, electromagnetic radiators like magnetic resonance imaging systems and other medical devices, large moving objects like supply carts and delivery carts can all impact RF-based systems. Many of these factors, like electromagnetic radiators and large moving objects, are variable and can also affect positioning reliability. To improve the reliability of the system, combine different technologies.

7.6.3.3 PLANNING, DEPLOYMENT, AND OPERATIONAL CONSIDERATIONS

The cost and complexity of planning, designing, and deploying a positioning system vary greatly. Planning typically entails a site survey of varying depth, collecting data regarding the customer requirements, and securing floor plans. A clear idea of exactly what steps are required to scope, plan, and design the RTLS, the responsible parties for each step, and the time required is essential. Some RTLS vendors may "guesstimate" on certain portions of the planning. Be definite, there is a clear agreement on who would be responsible in the event that guesstimates result in unanticipated costs. Deployment is the installation, configuration, and testing of the system. Floor plans are entered into the system, along with text descriptions of specific locations so that the RTLS can return both an icon on a map or a string of text to point out the location of a tag. Receivers are installed and RTLS positioning accuracy tested against requirements. The Tag registration process associates a tag with a specific asset, staff, or patient. Based on the application, rules may need to be configured for workflows the RTLS is intended to automate. Messaging configuration is also required to indicate who, when, and how an individual (or role) is to be notified when the RTLS generates an alert/alarm based on defined workflow rules.

Operational considerations: Include the ease or difficulty of maintaining tag registration data on the item associated with the tag. Creating new or changing existing rules and notifications are a significant part of RTLS management. Maintenance may need for tags and receivers. Software and physical reconfiguration will make when the hospital physical plant changes due to new construction, renovation, or simply moving units and departments. Depending on the degree of physical plant changes, the entire planning and deployment cycle may need to be repeated.

Costs: The two major cost components are first cost measures about actual labor, hardware, and software for proposed system; second component is how the RTLS system is marketed and sells to the hospitals. Other cost considerations based on the number of the tags, technology, number of receivers required to get the desired positioning accuracy and also other components such as software and servers or appliances. A significant cost variable is receiver installation; receivers have need of power and a network connection. Some hospital environment needed customized RTLS system, this increase the cost. The purchase of the system may be rental, own purchase, or lease from third party. Also include the service cost for consideration.

7.7   CONCLUSION[7]

RTLS market in healthcare sector hits \$2.92 billion by 2023 at a CAGR of 19.2%, the mentioned hit is based on the market research report by Industry ARC titled "Real Time Location Systems Market: By Component (Tags, Software, Sensor, Services) Technology (Active RFID [Bluetooth, WI- Fi, Others], Passive RFID) Tracking Use Cases (Patient, Asset, Personnel, Temperature Monitoring)-Forecast (2018–2023)."

Key players of Real-Time Location Systems in Healthcare Market:

Some of the top players in global market are IBM, GE Healthcare, Zebra technologies, Stanely healthcare, etc. They are strives to work hard to gain the position in RTLS market, this leads improve in product capabilities with the choice of product selection.

Example:

Zebra Technologies captured Motorola Solutions' Enterprise division. This getting hold of has enhanced Zebra Technologies mobile solutions product portfolio. This getting hold of is planned to leverage Motorola Solutions' asset tracking services to augment the RTLS portfolio.

KEYWORDS

- **RTLS**
- **RFID**
- **sensor**
- **tracking**
- **SCM**

REFERENCES

1. www.decawave.com
2. www.qburst.com
3. www.advancedmobilegroup.com
4. www.semiconductorstore.com
5. Snowday, H. T. Understanding RTLS In Healthcare. *Healthcare Technol.* [Online], 1. www.healthcaretechnologyonline.com
6. Hegli, R.; Gee, T. Real-time Location Systems (RTLS) in Healthcare. www.skytron.us
7. www.digitaljournal.com

CHAPTER 8

Data Analytics and Its Applications in Cyber-Physical Systems

A. SHEIK ABDULLAH[1*], R. PARKAVI[1], T. SARANYA[1],
P. PRIYADHARSHINI[1], and ARIF ANSARI[2]

[1]*Thiagarajar College of Engineering, Madurai, Tamil Nadu, India*

[2]*Marshall School of Business, University of Southern California, Los Angeles, CA 90089, USA*

Corresponding author. E-mail: aa.sheikabdullah@gmail.com

ABSTRACT

The development in information and communication technology (ICT) depends upon the future trend and pattern generated by data analytics and its models. The industrial revolution depends on the models and patterns that have been reliably generated by machine learning and artificial intelligence techniques. However, the domain has still been going in advancement with methods in deep learning and analytical tools with mathematical models, since mathematical model plays a significant role in supporting the proof-of-evaluation for the machine learning and deep learning practices that have been deployed for the industrial needs.

Based on the data generated by stream applications, the applicability of the analytical model varies in terms of parameters, platform, model selection, and visual data exploration. This has to be considered as an important phenomenon in CPS because it may lead to some erroneous evaluation with regard to the industrial needs and its applicability. This chapter provides a detailed work which explicitly reveals the impact of data analytics and its types with its application toward CPS, and rapidly developing technology that helps in transforming the needs of Industry 4.0 and the society.

The following are the key features of the chapter:

- Exploring all the types of analytics and its applications
- Showcasing use cases for the domain of data analytics and CPS
- Implementation results exhibiting the mathematical models across analytics
- Applicability of machine learning algorithms in prediction
- Future research direction

With this consideration, the development of models with regard to CPS can be made specifically with mathematical formulations. The industrial revolution is targeted with a focus on new data models and its applicability to CPS. Meanwhile, the impact and consideration of data analytics play a significant role in prediction and data classification. This chapter explores all the possibilities and framework that can indulge data analytics with CPS in accordance with Industry 4.0 standards.

8.1 DATA ANALYTICS AN OVERVIEW

Every day a vast amount of data gets generated which are in various forms and are all help to get valuable information from them. The unprocessed information is in raw format, need to convert it into useful information where the process called analytics. Analytics is nothing but a continuous processing of data with the help of analytics tools or the efficient application of the algorithms and methods to process the data to get information from them. This information helps us to make better and meaningful decisions to meet the business requirements.

Data are generated through our daily process and these data are in various formats. Generally, the data are classified into qualitative and quantitative data. In qualitative data, the quality of the data is gets measured whereas in the quantitative data, the quantity or the number of data matters. The data can also be divided into two forms such as continuous and categorical. In the continuous form, the data follow a sequence order. In the categorical form, the data are further classified into three forms which are nominal, ordinal, and binary. In nominal, the data are not in a meaningful sequence and in the ordinal, data exist in a meaningful order. In the binary, the data can only be in two forms alone. Analytics process starts its process from the raw format which is then moved to the data selection where the needed data are alone selected and moved to the preprocessing step where the data cleaning process takes place. It includes the missing tuples, outliers detection, the removal of duplicate values, and the numeric conversion of data.

After this preprocessing step, the data are transformed into another format. Finally, the interpretation and evaluation process is done, which is the outcome of the analytics process. Interpretation helps to interpret the result of the preprocessed data with the help of the facts, knowledge of the process, and the use of the appropriate methods. It can be applied to both qualitative and quantitative data formats. Evaluation of data is done with the help of various metrics and measuring parameters, which are applied on the data during the process of analytics. Various analytic tools are available to perform the intended function on the data. The tools range from the excel to the advanced big data analytic tool such as Hadoop. Some of the tools are R-programming, Rapid Miner, Tableau Public, SAS, and Splunk.

These tool helps to address the need of analytics in various fields, which also required large amount of data processing to provide the meaningful decision. The current trends of data analytics include a variety of technologies and fields. This analytic process helps to increase the accuracy of the data and also helps to increase the processing power of the data. It consists of Internet of Things (IoT), Machine Learning, Graph Analytics, Artificial Intelligence, and Augmented Reality and Cyber-Physical Systems, etc.

8.2 TYPES OF ANALYTICS AND ITS APPLICATIONS

Analytic process is of various types which are descriptive analytics, predictive analytics, diagnostic analytics, and prescriptive analytics. Descriptive analytics is the most frequently used process where it performs analytics in a way that is easily understandable to humans and is in detailed described way. This type of analytics also helps to find the wrong things from the previous year's data, which are very useful for future process. Predictive analytics helps to predict the future data with the help of the current events. This type of prediction is done with the help of the probability of happening the events in the future process. Here, the accuracy of the predicted value depends on the quality and selection of data for the analytic process. Diagnostic analytics is done with the history of the processing data which are used to perform the analytic process.

This process requires a deeper analysis of the data which are further used for the diagnosis of the data. So for this type of analytics, there is a need of detailed information about the processing data for the better processing of data. Prescriptive analytics is generally done to prescribe the solutions and details for the particular problem that the model tries to solve. This analytics helps to provide the better solutions for the raised problems and also make the process more reliable to use and the process called as solution-oriented analytic process. These analytics types are used based on the need and the process is completely applicable to the related problems, which are dependent on the user's needs. These analytics processes can be applied to a variety of problems and situations. Some of the most frequently used applications are recommendation systems, image and speech recognition, gaming, airplane travel planning, risk and fraud detection, self-care driving, customer interactions, and also in the field of robotics.

8.2.1   RECOMMENDATION SYSTEMS

In today's world, there are large number of recommendation and advertisements based on our previous searches and interests in the webpage. This type of recommendation is highly achieved with the help of these data analytics processes where the prediction of the future purchases are guessed correctly which are shown on the desktop screen. These are highly profitable for their business which induces the customers to buy to through these constant recommendations.

8.2.2   IMAGE AND SPEECH RECOGNITION

In social media, when we try to upload a picture the tagging option is automatically popup which is purely based on the image recognition which is achieved with the help of data analytics. In Pinterest, the image search option is available which helps to scan the image and results similar image from the website and it can be done easily with the help of data analytics. In Google home, Siri and Alexia are able to work and process data based on the voice and speech recognition which are highly depending on the data analytics process. The accuracy of the queried results is highly dependent on data analytics methods.

8.2.3   GAMING

Gaming is another area where data analytics is already in vast use. The gaming field uses machine learning algorithms and data analytics for their efficient working process and also it can be used in the motion gaming where your moments trigger the actions on the game which are all under the help of data analytics.

8.2.4   AIRPLANE TRAVEL PLANNING

Traveling in today's world population needs a helping hand for the processing of data to make the process easier and customer friendly. In airplane travel planning, the prediction of flight delay and any other suggestion related for booking seats all are works under the data analytics process. This data analytics process helps to precisely locate and halt the flights which avoid the unnecessary collisions and accidents in the route.

8.2.5   RISK AND FRAUD DETECTION

Advanced detection of risks and fraud is all done with the help of analytics. For example in the intrusion detection system, the identification of true positives is done with the help of analytic process which reduces the false negatives and helps to minimize the risks in the systems. Along with the detection, the recommended solutions are also being suggested with the help of the diagnostic analysis process.

8.2.6   ROBOTICS

In robotics, a huge development is made with the deep learning and neural networking process which needs the base of data analytics. Here, the combination of sensory and cognitive functions mainly depends on the quick processing and analysis of data to make fast and correct decisions at the same time with the help of these data analytics techniques.

As the need of analytics is huge in the today's world and its applications cannot be limited in small areas. This analytics process is an advantage to the various fields which include numerous algorithms and methodologies.

8.3   DATA ANALYTICS AND PROCESSING PLATFORMS OF CPS

Currently, Machines and devices are connected as a collaborative community in Industry 4.0 and play a vital role in industrial production. This encourages a computerized manufacturing in a way to make decentralized decisions. The challenge of cyber-physical system is to bring out more meaningful and intelligent insight and yields optimal decisions from the industry data because such a revolution in an industry will lead to generate a huge amount of data. With the help of big data, CPS will collect, store, and make analytics in a real-time. This will improve the efficiency of industrial production.[10]

Automated decision making is one of the features of CPS. The CPS has several objects interacting with one another which in turn produce massive amount of data in order to extract the hidden pattern to make optimal decision and to enhance the quality of service. To bring big data environment into CPS, we need a resilient network. Big data processing platforms are varying for different system such as Hadoop for batch processing system, Storm for stream processing system, and cloud computing storage model.[11,12]

8.3.1   PROCESSING PLATFORM FOR CPS

Hadoop is an open-source software framework for storing a massive amount of data and running the application which consists a cluster of commodity hardware. For handling such a quantity of data, it has a core element called Hadoop Distributed File System (HDFS) and MapReduce. This enables a distributed processing of a large amount of data and it can be processed in a reliable, scalable, and fault tolerance.

Apache storm is an open-source that processes real-time streaming data at a very high speed when we compared it with Hadoop. It is also used to create a complex event processing system, which is classified as computing and testing. Cloud computing is highly needed to process and analyze the distributed and enormous quantity of data. The cloud supports reliable architecture to perform analytics for CPS data on a big stream of data such as extraction and aggregation. Cloud computing for CPS big data has a better effect on the real-time needs. The cloud enables us to perform parallel computations on the CPS data items thereby, we can achieve speed. According to the cloud security alliance (CSA), most of the enterprises still not yet move to the cloud due to the security problem.

The design of CPS structure should consist of connectivity which ensures the real-time data streaming and there is a need for data analytics in an intelligent way. The 5c structure proposed by Lee et al.[3] that show how to construct a CPS system from data acquisition, such as connection, conversion, cyber, cognition, and configuration level.

Connection: This is responsible for data acquisition and transfer to the central server.

Conversions: This is responsible for discovering meaningful insight. It converts the data to information with the help of intelligent algorithms and data mining techniques.

Cyber: This level consists of more enormous information which is needed to take intelligent decision and acts as a central hub. The flow of data mining and cyber-physical system is depicted in Figure 8.1.

Cognition: This creates knowledge of the monitored system by implementing CPS.

Configuration: It is the feedback from cyberspace to physical space. This is responsible for the machine to take self-configuration and self-adaptive by act as a supervisory control.

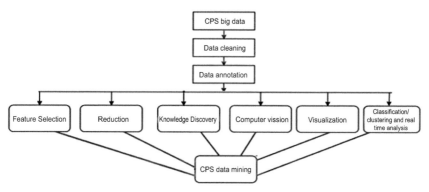

FIGURE 8.1 Data mining and CPS.

Atat et al.[1] presented an overview of data collections, storage, access, processing, and analysis for CPS taxonomy. From Figure 8.1, CPS data mining: Extracting useful information from the CPS data collected from the various sensor or device. Before mining CPS data, we need to apply some processing steps such as features selection, preprocessing, and transformation of data. The reduction is the technique used to reduce the dimensionality of the data this can be done by Principle Component Analysis (PCA). Knowledge discovery is highly used in different CPS scenarios to find out the unknown correlations in CPS data. Classification and Clustering are the two different mining techniques which make the CPS smarter.

Lee et al.[2] address the trends of service transformation in big data environment and the smart predictive tools to manage big data. They stated that health awareness analytics of machine with the self-learning Knowledge base. If the health information of a machine is a need then the Knowledge base will generate the necessary information and the prediction algorithms. For creating clusters autonomously for different working regimes and machine conditions, an unsupervised algorithm is highly used such as Self-Organizing Map (SOM) and Gaussian Mixture Model (GMM).

Rehman et al.[6] have proposed a framework to handle big data for cyber-physical system and this framework considers the possible solution such as standardization, cloud computing, online, and data stream learning to analyze and process a CPS data. Their proposed framework addresses a challenge on big data for CPS such as real-time, infrastructure, data quality, and security.

8.4 MACHINE LEARNING TECHNIQUES IN PREDICTION

Machine learning is a concept comes from artificial intelligent. Without any explicit program or instruction, the computer machine or a computerized device will perform a specific task efficiently. This is a study of algorithm and statistical model.

The machine learning algorithms for predictions are:

- Supervised machine learning.
- Unsupervised machine learning.
- Reinforcement machine learning.

8.4.1 SUPERVISED MACHINE LEARNING

In supervised learning, the machine learns with the help of labeled data for training. There are two types of supervised learning such as Classification and Regression. When the output variable is in the form of real values then regression is used to predict. If the output variable is in the form of categorical then the classification is used to predict.

8.4.2 UNSUPERVISED MACHINE LEARNING

In unsupervised machine learning, the machine learns without the help of labeled dataset. There are two types of unsupervised learning such as clustering, association and dimensionality reduction.

8.4.3 REINFORCEMENT LEARNING

In reinforcement learning, the software agent plays a vital role to decide the next action based on their current state behavior. This learns by the trial and error method.

Sargolzaei et al.[5] proposed the neural network for fault detection in vehicular cyber-physical system. The fault detection technique is applied to detect and track fault data injection attacks on the cooperative adaptive cruise control layer of a platoon of connected vehicles in real time. Bezzo et al.[4] has proposed a reach ability-based approach and a Bayesian Inverse Reinforcement Learning Techniques to predict a malicious intention in cyber-physical system under cyber-attack. The reachability is used to determine the set of possible states that the CPS may cover over a certain time horizon, because the input may be uncertain due to sensor noise. Then they apply the Inverse Reinforcement Learning in order to identify the intention of the attack.

8.5 USE CASES DEPICTING CPS AND ITS MODELING ENVIRONMENT

Cyber-Physical Production Systems (CPPS) view as a replacement of the forthcoming Smart factory environment systems due to their mountable and flexible nature. CPPS will help in the integration, adaptation, and replacement of production units in terms of any distractions and disappointments. With the recent development of digital in this physical world, the next generation of CPSS or Smart factories will be most important. In this today world, products are flattering multifaceted and their lifecycle reduces, due to the usage of simulation tools optimization and acceleration of all stages of lifecycle get more attention.

The main objective of the simulation tools is to up keeping re-engineering and decision-making process in the production unit and to assess whether internal and external controls are working properly and if any fault occurs how to overcome the fault within a particular time period. To model and simulate the Smart Factories, ICT and Automation Technology (AT) merge together to address the challenges and problems faced by the consumer and market. With the recent development of digital tools, it can support both the management and the production unit. The data attainment part of the CPS collects existing standards and combines them into a single extensible CPS. Consequently, the CPS data structure turns as a container model that preserves semantic links between different standard descriptions and scope for stipulating asset, pertinent sub-models, and behavioral models. The metamodel comprises of different sections, it contains information related to the CPS such as vendor, version, type, and owner.

The first use case depicting CPS and its modeling environment is the most competitive, advanced, and complex industrial sectors, the world's largest carmaker in Europe: The Automotive Production and Body Shop of Volkswagen. A framework is designed in a way to enable a continuous connection of vendor-specific CPS, which covers the way to segment their simulation-based models and outcomes of this model maintains multidisciplinary simulation by connecting collaboration and organization between CPS models and successive simulations as shown in Figure 8.2. This architecture framework contains a three-tier approach, disconnected into a presentation, submission, and data tier. This framework is implemented in the Automotive Production of Volkswagen and the implementation part follows the principles of Service Oriented Architecture (SOA).[6]

CPS is a combination of physical procedures with computation and communication, CPS will be able to enhance more intellectual to social life. Wireless Sensor Networks (WSN) can be a vigorous part of CPS as a robust identifying ability is one of the main dynamic features for CPS applications. With the recent development and advancement of WSN, Medical Sensors, and Cloud Computing, CPS will control healthcare applications such as in-hospital and in-home patient care. CPS will remotely access the patient's conditions and provide necessary consultation remotely.

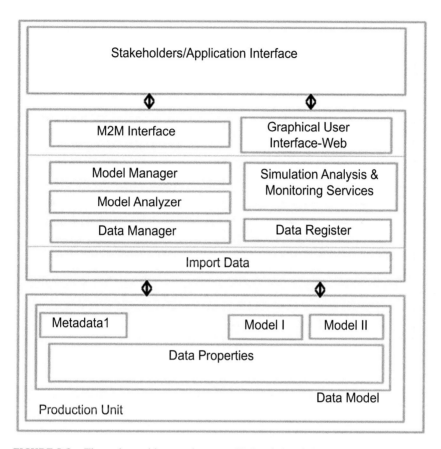

FIGURE 8.2 Three-tier architecture between CPS and simulation production life cycle.

The data collected are stored in a server and it authorized only clinicians due to security reasons. When compared to other applications, CPS architectural design in healthcare application is designed with more concern to ensure data security. Another thing to consider while designing a CPS database management system is to be designed to store and manage thousands of patients records collected from multiple medical sensors. Other than that CPS requires large computing resources for taking actions remotely based on patient illness or disease.

CPS in healthcare includes application, architecture, sensing, data management, computation, communication, security, and control as shown in Figure 8.3. Application is split into two: Assisted includes in-home and elderly living people and Controlled application includes hospital and intensive care patients. Architecture is split into three: Infrastructure includes Server-based and Cloud-based infrastructures, data requirement includes light and heavy data, Composition system includes user-defined system and automatic system. Sensing is split into three categories: Sensor type includes homogeneous sensor and heterogeneous sensor, Sensing method includes active and passive methods and Parameters are Single and Multiple parameters.

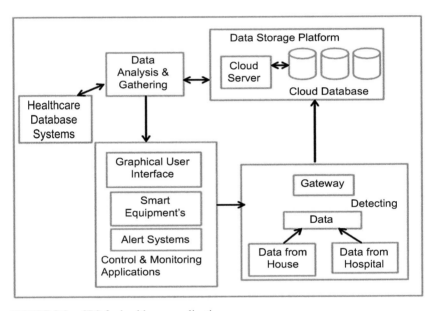

FIGURE 8.3   CPS for healthcare applications.

Data Management System comprises of Data Integration, Data Storage, and Data Processing, Data Integration includes individual and combined data integration, Data Storage includes Central and Distributed data storage and Data Processing includes In-network, Base station, and Cloud. Computation is split into two: Modeling and Monitoring, Modeling includes user-defined static and static stimulation, dynamic Prediction and Just-in-Time Dynamic, Monitoring includes Health status, Daily living, and Intensive care. Communication is split into two: Scheduling and Protocol. Scheduling includes Combined Scheduling and Sub Scheduling, Protocol includes Centralized Protocol, Hierarchical Protocol, and Decentralized Protocol. Security is split into two: Privacy and Encryption, Privacy includes Application Level Privacy and Data Level Privacy; Encryption includes User Level Encryption and Network Level Encryption. Control is split into Mechanism and Decision Making, Mechanism includes manual mechanism and automatic mechanism; Decision Making includes Single parameter decision making and multiple parameters decision making.[7]

8.6 EXPLORATORY DATA ANALYSIS FOR PREDICTION

With the emerging Industry 4.0 technology, CPPS activates a standard move from vivid to narrow maintenance management approach. There are numerous approaches already existed, still, there are some challenges in reconsidering maintenance management in the Industry 4.0 context. In terms of production management and planning, prescriptive maintenance model (PriMa) is introduced.[8] This prescriptive maintenance model consists of four layers: Data Management Layer, Predictive Data analytic Toolbox Layer, Recommender and Decision Support Dashboard Layer and Overarching Layer. PriMa has two useful competencies: the first one is, it can be able to efficiently process huge volume of data retrieved from multiple data sources and the second one is, it can be able to provide decisions and references for refining and improving the maintenance tactics linked with production planning and control (PPC) systems.

To implement smart and knowledge-based solutions in the smart factories; incorporating maintenance approach with functional capabilities is the only way. A recent report reveals that Industries predictive and prescriptive maintenance includes equipments, types of machinery, and physical assets it will be around 79% in the upcoming 3 years. Whereas maintenance is the major part of the production industries and also it is the most critical part of the overall production system. With the recent technological developments in the area of digital, automate, and intelligent in the manufacturing industry, it is difficult to implement these recent technologies to predictive maintenance in terms of finding errors and detection of failure. Prediction of failure is the maximum significant part of the maintenance management to prevent and decide tenacious problems like unavailability of resources, instability of process, and inefficiency of resources. CPPS environment retaining detecting and computational technologies and present data-driven procedures in the maintenance and activate rethinking maintenance methods through ongoing investigation of innovative knowledge and inspirational as well as conserving existing knowledge.

There are many similarities between original maintenance methods improved by sensing and computational technologies, but there is some lack of accord in qualifying the maintenance approach in Industry 4.0. To address this, PriMa includes five different functional capabilities for describing maintenance in the Smart factories, which are Prediction Capability, Optimization Capability, Adaptation Capability, Learnability and Capability of Intelligent actions and self-direction.

In terms of Industrial application of PriMa, it relies on supervised learning and reasoning, it also includes data analysts and knowledge engineers work together with domain and business experts. Industrial use-case considers here is an international manufacturer of gearboxes and engines for the automotive sector. The main objective of this manufacturing company is discovering possibilities to employ data-driven approaches and develop a new knowledge-based maintenance plan. Cross-Industry Standard Process for Data Mining (CRISP-DM) consists of six phrases: Business Understanding, Data Understanding, Data Preparation, Modeling, Evaluation, and Deployment. Based on this, data are collected, preprocessing and finally, a decision is taken. The first step is data transformation, for this, text mining methods are used, the second step is data identification, to find the correlation between features, the third step is framing rules based on the first and second step, based on these rules prescriptive maintenance decision is taken.

In recent studies, manufacturing industries produce a large volume of data when compared to other sector or industries, so far most of these are not connected by companies. One example of this, the oil and gas company rejects 99% of their data even before the decision is taken by the experts. Considerably, these kinds of things are highlighted even with the introduction of smart manufacturing and its combined enterprise technologies. Smart factories mainly focused on control optimization and intelligence, once a better intelligence can be attained by relating different neighboring systems that directly impact the performance of machinery. Today's factory automation is driven by two leading and interoperable developments they are Industry 4.0 and Industrial IoT (IIoT). Industry 4.0 represents the business objectives, intelligent algorithms, analytics, predictive technologies, and cyber-physical systems. IIoT represents an implementation of the Information Technology domain and IoT and includes sensors, actuators, control systems, communication, data analytics, and security. Industry 4.0 and IIoT provide jargons like "Data is everywhere,"

"Instrumentation Everywhere," and "Connecting Everywhere" are the principles that validate the statistic that recent industries have a huge volume of data to deal.

The use-case considers here is Alarm and Event and Exploratory Data Analysis (EDA) is used for data analysis is shown in Figure 8.4. The main purpose of EDA is to find useful information from the data collected and also find patterns to further analysis. EDA is essential to understand the data at first. Data are representing either in categorical or continuous form. If it is categorical data EDA mainly use visualization techniques and graphical methods for representing the relationship, patterns, and categories among variables. Graphical methods are recently used in categorical types of data.

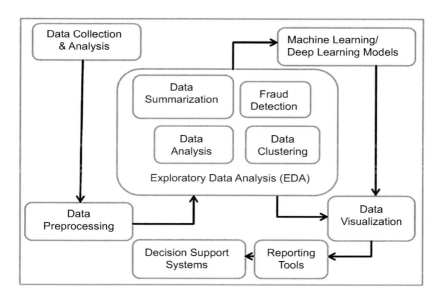

FIGURE 8.4   Data science workflow-EFA.

In EDA, it is important to launch a possible relationship among variables during analysis. multiple correspondence analysis (MCA) is used to brief and visualize the table of data which contains more than two categorical data. MCA is an extension of Correspondence Analysis (CA) and it is a part of the Principal Component (PC), its main purpose is to brief and visualize the most significant data in a multivariate dataset.[9]

8.7   PRIVACY ISSUES

Cyber-physical system protects to the personal information and it can be ensure the authorized person privacy and works. The personal information likes name, addresses, telephone, email addresses, professional certificate information, and personal financial information. The CPS security and privacy characteristic are CPS domains, attacks, research trends, and security level implementation. Attacks are mostly affecting to high integrity of the system device. It should be directly affect to the physical part and cyber component. Physical attack is consisting of focus to physical element for CPS and cyber attack is considered to gain access to communication network. Data analysis is huge amount of quantity of data to be analyzed for at a time. These data are unstructured and various different type of dataset such as video and image archives, sensor, social media, and medical data. Big data generate to the connection of more device that connected device called as IoT. The IOT is produced the large number of data which is transfer to the information from one device to another device. So physical device privacy is need to the data analytics techniques. Security and privacy are important concerns of cyber-physical system. Privacy is the main focusing point for front and center for system designer.

To improve privacy in CPS first it can be identify the vulnerabilities of CPS. Find the hardware cyber based thread and attacks. Prevention is a recover to the attack for and improves the security futures. Detection is analyzed for attacking virus or malicious and recovered to loosing information. Protection is described to encrypt the information for communication.

8.7.1   ATTACKS

8.7.1.1   EAVESDROPPING IN CPS

Eavesdropping is injected to attack the wireless communication through connecting device. These attacks to be analyzed for two different ways one is device emission analysis and another one is power analysis. Device emission analysis is performing to electromagnetic emission device and RF radiation in communication.

8.7.1.2 DENIAL OF SERVICE

DOS attack is active attack for ICMP based flooding. It can be attacked to target for availability of physical system. DOS attack highly affected and damage to cyber-physical system. DOS targets include to CPU memory, exploiting time, and congesting network power.

8.7.1.3 DDOS ATTACK

DDOS attacks have a SYN flooding attacks. It is sent to multiple of request to CPS and it is damaged for physical system.

8.7.1.4 COMPROMISED KEY ATTACK

A secret code is generated to attacker and inject to physical system. Information is considered to the key format and information to be accessed used to key. It is compromised to physical device and gain to be accessed.

8.8 REINFORCEMENT LEARNING AND DEEP NEURAL NETWORK FOR AUTONOMOUS DRIVING

Reinforcement learning is an AI can be used to train the machines and it is interacting with environment. Reinforcement is not to be successfully applied in automatic application. The learning of robot car requires to high skill. This method has focus on human detection. Autonomous driving agent can be categorized such as recognition, prediction, and planning.

8.8.1 RECOGNITION

Recognition can be identifying the component for nearby surrounding environment. Most of the recognition method solved to using the deep learning algorithm. In which that recognitions are human level object detection and classification method. More complex to learn the feature of raw input data but deep learning is easy to learn the complex of raw data. Convolutional neural network is the new trending successful method for deep learning algorithm. Convolutional neural network is a one kind of artificial or deep neural network. It is specifically designed to process pixel of data.

In this algorithm, hardware and software are act as human brain. It should be managed and stored for high level data. Artificial neural network used for processing the Visualize image analyzing in which those neural networks mostly apply to image visualization. Convolution neural network is consisting of three main layers like input, output, and hidden layer. It will be trained to image classification. There are three ways of training to understand the basic forms. First understand to the edges of image, second learn to structure and color and a final point is classify the image. There are several layers in CNN in which input layer takes images. Convolution layer is learning by subsampling then pooling layers stores the samples from the image convoluted. Finally, fully connected layer collects all the samples it should be find the proper output. This algorithm successfully applied to vehicle detection for autonomous driving.

8.8.2 PREDICTION

Prediction method can not be enough for a reinforcement autonomous driving environment. Prediction method tracks object for near environment and in which that predict the future. The future information is must be important to deep learning process. Such that recurrent neural network is used to end-to-end system. In this algorithm improve the object detection and tracking performance.

8.8.3 PLANNING

Planning is a difficult task for automation driving. It is very difficult to understand the reorganization and prediction performance level. Learn to future action such that removes to unwanted penalties. Planning is a decision making process of deep learning algorithm.

Reinforcement is control the time taking task. It does perform to human level of control. Reinforcement learning (RL) is considering to learning stage and Deep Learning (DL) consider to learning stage algorithm. This algorithm mixed to use in reinforcement method that is also known as Recurrent Neural Network (RNN) algorithm.

8.8.4 DEEP NEURAL NETWORK

Deep neural network (DNN) based vehicles are categorized into two types like feed forward CNN and recurrent neural network. The deep neural network should be presented in convolution layer. DNN consisting of individual sequenced unit, these units are called as neutrons. Neutrons are connected to different layers that neutrons have same weight. In this, neutrons are applied to the nonlinear of function activation. DNN training process involved to the labeled of training dataset. DNN once trained it is automatically predicted for without future interaction.

8.9 RESEARCH MOTIVATIONS AND RECENT TRENDS

Recently, manufacturing industries adapted Industry 4.0 environment and cyber-physical systems, and data analytics plays a role in processing the data gather from the sensors or other data sources. Based on the data collected, decision will be taken to improve the production system and to tackle errors in case of system failures. Research motivation in CPS is to fully integrate the production system in manufacturing industries and automated all the process, future research expects more works are to be done in healthcare applications.

8.10 CONCLUSION

Cyber-physical systems have its applications toward a wide range of diversified domains such as Sensor Navigation, Medical Informatics, Weather, and Security Constructs and so on. The incorporation of CPS in the domain of data analytics provides the pathway of securing and analyzing the data with algorithmic constructs and its preprocessing strategies. This chapter completely focuses on the applications and techniques that have been concerned with data analytics for CPS. Algorithmic development strategies and their issues have also been addressed with regard to the computing models and its processing platforms. The incorporation of data analytics plays a significant role in modeling CPS and their applications. Hence, this chapter could benefit the readers in finding out a path for analyzing and modeling data analytics for cyber-physical systems.

KEYWORDS

- data analytics
- big data
- machine learning
- cyber-physical systems
- industrial revolution

REFERENCES

1. Atat, R.; Liu, L.; Wu, J.; Li, G.; Ye, C.; Yang, Y. Big Data Meet Cyber-physical Systems: A Panoramic Survey. *IEEE Access* **2018**, *6*, 73603–73636.
2. Lee, J.; Kao, H.-A.; Yang, S. Service Innovation and Smart Analytics for Industry 4.0 and Big Data Environment. *Procedia Cirp* **2014**, *16*, 3–8.
3. Lee, J.; Ardakani, H. D.; Yang, S.; Bagheri, B. Industrial Big Data Analytics and Cyber-physical Systems for Future Maintenance and Service Innovation. *Procedia Cirp* **2015**, *38*, 3–7.
4. Bezzo, N. Predicting Malicious Intention in CPS under Cyber-attack. In *2018 ACM/ IEEE 9th International Conference on Cyber-Physical Systems (ICCPS)*; IEEE, 2018; pp 351–352.
5. Sargolzaei, A.; Crane, C. D.; Abbaspour, A.; Noei, S. A Machine Learning Approach for Fault Detection in Vehicular Cyber-physical Systems. In *2016 15th IEEE International Conference on Machine Learning and Applications (ICMLA)*; IEEE, 2016; pp 636–640.
6. Weyer, S.; Meyer, T.; Ohmer, M.; DetlefZuhlke, D. G. Future Modeling and Simulation of CPS-based Factories: An Example from the Automotive Industry. *IFAC-PapersOnLine* **2016**, *49-31*, 97–102.
7. Haque, S. A.; Aziz, S. M.; Rahman, M. Review of Cyber-Physical System in Health-care. *Int. J. Dist. Sens. Netw.* **2014**, Article ID 217415, 20 pages; http://dx.doi.org/ 10.1155/2014/217415
8. Ansari, F.; Glawar, R.; Nemeth, T. PriMa: A Prescriptive Maintenance Model for Cyber-physical Production Systems. *Int. J. Comput. Integr. Manuf.* **2019**, *32* (4–5), 482–503; DOI: 10.1080/0951192X.2019.1571236
9. Bezerra, A.; Silva, I.; Guedes, L. A.; Silva, D.; Leitao, G.; Saito, K. Extracting value from industrial alarms and events: a data-driven approach based on exploratory data analysis. *Sensors* **2019**, *19* (12), 2772; https://doi.org/10.3390/s19122772
10. Abirami, A. M.; Sheik Abdullah, A.; Selvakumar, S. Sentiment Analysis. *Handbook of Research on Advanced Data Mining Techniques and Applications for Business Intelligence*; IGI Global. ISBN: 978-1-5225-2031-3; DOI: 10.4018/978-1-5225-2031-3.ch009

11. Sheik Abdullah, A.; Selvakumar, S.; Abirami, A. M. An Introduction to Data Analytics: Its Types and Its Applications. *Handbook of Research on Advanced Data Mining Techniques and Applications for Business Intelligence*; IGI Global. ISBN: 978-1-5225-2031-3; DOI: 10.4018/978-1-5225-2031-3.ch001.

12. Sheik Abdullah, A.; Suganya, R.; Selvakumar, S.; Rajaram, S. Data Classification: Its Techniques and Big Data. *Handbook of Research on Advanced Data Mining Techniques and Applications for Business Intelligence*; IGI Global; ISBN: 978-1-5225-2031-3 eISBN: 978-1-5225-2032-0; DOI: 10.4018/978-1-5225-2031-3.ch003.

CHAPTER 9

Dual-Axis Solar Tracking and Monitoring of Solar Panel Using Internet of Things

VIJAYAN SUMATHI[1*], J. KANAGARAJ[2],
SIVA SARATH CHANDRA REDDY[1], SAI SUJITH KANKIPATI[1],
ADARSH VIDAVALURU[1], and UMASHANKAR SUBRAMANIAM[3]

1School of Electrical Engineering ,VIT Chennai, Tamil Nadu, India

2CSIR-CLRI, Adyar Chennai 20, Tamil Nadu, India

3Prince Sultan University, Riyadh, Saudi Arabia

Corresponding author. E-mail: vsumathi@vit.ac.in

ABSTRACT

This chapter presents the design, development, testing, and implementation of a dual-axes tracker. The study also presents the technique of increasing the efficiency of the solar panel by changing the orientation with respect to the sun's position using dual-axes tracker. Light-dependent resistors are mounted on the solar panel to track the movement of the panel. The movement of the panel is also controlled by two 12 V DC motors for each of the axis. Arduino microcontroller is used as the interfacing medium to control the motors. In order to determine the direction of rotation of motors, relay circuits are also used as motor drivers. The Wi-Fi module is used to monitor the entire process through the IOT. The results are evaluated throughout the day and are evaluated with and without the tracker. The performance of the developed system was experimented and compared with the static solar tracking system. This work further demonstrates that the dual-axis solar tracking system has a higher power generation compared to static panels. The experiments show that employing dual-axis tracking system under sunny conditions has resulted in an increase of the voltage generated by around 10% in the morning and evening but with no significant change during the afternoon. The proposed design is achieved with low power consumption, high accuracy, and low cost.

9.1   INTRODUCTION

It was in the early 19th century when the first steam engine came into existence, was the period when man got to know about the importance of fuel. Fuel might not only be steam, as in steam engine, but also coal, petroleum, and natural gas to name a few. By the end of the 19th century and the beginning of the 20th century, the use of coal and petroleum as core fuels was widely regarded as the ultimate sources of industrial revolution and development. But soon, scientists, mathematicians, and geologists inferred that these fuels were limited and it might take another hundreds of thousands of years for the Earth's biosphere to replenish them. Thus, these fuels are referred to as non-renewable resources.

Gradually, the exploration for a better alternative in place of these non-renewable resources other than steam led to the discovery of use of forces and elements of nature as fuels. Thus, hydro, geothermal, biomass, wind, tidal and solar energy gained importance as the time and technology advanced. These were called as the renewable sources of energy. These sources of energy are almost inexhaustible and easily restorable in nature.[1] Out of all these sources, solar energy is the most abundant and least explored form of energy in terms of technological advancement.

The average amount of solar energy that Earth receives in the form of solar radiation is 174,000 TW out of which 30% is reflected back into the space. The Earth receives about 1.36 MW of direct solar radiation per square meter on a clear sunny day out of which 50% is visible light, 45% is infrared and the remaining is other forms of electromagnetic radiation. Even harnessing a little of this massive amount of energy can bring about a big change in the way energy is utilized because harnessing energy from solar radiation is environment-friendly.

There are two ways in which solar energy can be harnessed.[2] The first method is the conversion of solar radiations into thermal energy and second the conversion of those radiations into electrical energy. Solar energy can be converted into electrical energy with the help of solar cells, also known as photovoltaic cells. These cells are made up of silicon and work on the principle of photovoltaic effect.[3] When a solar cell is exposed to sunlight, the electrons present in the cell absorb the radiations, attain energy enough to free themselves from the inter-atomic bonds due to the effect mentioned above. This causes the flow of electrons within the solar cells. This movement of electrons produces electricity. A grid of solar cells constitutes a solar panel.

Even though energy generation from solar cells is eco-friendly and noiseless,[4] the efficiency of a solar cell is pretty low. Various factors affect the efficiency of a solar cell like the weather, quality of the solar cells, spatial orientation of the cells in the panel and the position of the panel with respect to the sun's position. The position of the panel can be changed manually or automatically by the use of a solar tracker which works on the principle of Maximum Power Point Tracking.

This chapter deals with the orientation of solar panel in dual axes with the help of automatic tracking of solar radiation.[5] This results in a significant increase in the efficiency of the energy conversion and thereby decreasing the total production costs of solar energy. The chapter also proposes a method of implementing the concept of IoT to monitor the performance of the solar panel during various duration of the day.

9.2 PROPOSED METHOD

The proposed method consists of LDR sensors mounted on the solar panel. When one of the sensors gives more output than the other sensor it is detected by the Arduino microcontroller which in turn rotates the wiper motor along the axis in which the LDR sensors are connected, such that the panel aligns itself with respect to the position of the sun. The LDR sensors detect equal intensities of solar radiation.[6]

The spur gear system and the Slider Crank mechanism help in the smooth movement of the solar panel in the desired angle and direction. When the voltage sensor is connected to the panel, it senses the voltage and sends the signal as an output to the analog pins of the Arduino. The signal received from the voltage sensor is converted into a digital value with the help of a predefined equation. The converted digital value of the voltage sensed is sent to the IoT platform with the help of a Wi-Fi module which helps in interfacing the Arduino with the internet. The process of sending the digital voltage values to the IoT platform is executed with the help of a C code downloaded in the other Arduino microcontroller.

9.3   DESIGN SPECIFICATIONS

Components	Specifications
Solar panel	5 W, 12 V
DC motor (wiper motor)	12 V, 120 rpm
Arduino microcontroller	Operating voltage – 5 V
ESP2866	Vin – 3.3 V
Voltage sensor module (B25)	Can measure voltage that is less than 25 V
LDR	Vin – 5 V
Voltage regulator circuit	Vin – 12 V, Vout – 5 V
Relay circuit	–
Connecting wires	–

9.3.1   ARDUINO UNO

Arduino UNO is a microcontroller based on the open source physical computing platform. It is a simple I/O board that implements the processing/wiring language. The microcontroller can be used to develop stand-alone interactive objects or can be connected to software on the computer. The programming is done using the software called Arduino IDE.[7]

9.3.2   VOLTAGE SENSOR

To measure the voltage, a voltage sensor B25 is used in the study. It gives an accurate voltage measurement for voltage up to a value of 25 V. These sensors are used for metering and measuring the overall power consumption of systems. The Arduino has the capability of measuring the voltage in the range between 0 and 5 V. Since the 12 V solar panel is used for voltage sensor. This sensor decreases the voltage by a factor of five and gives an analog output to the Arduino microcontroller that calculates the voltage obtained from the solar panel.

9.3.3   RELAY DRIVER

In each of the two channel relay circuit drivers, there are two relays dedicated to each direction of the wiper motor. The common terminals of the two relays are connected to the motor terminals and the NO terminals of both the relays are connected to the positive terminal of the 12 V battery while their NC terminals are connected to the negative terminal of the battery. The connections are shown in the Figure 9.1.

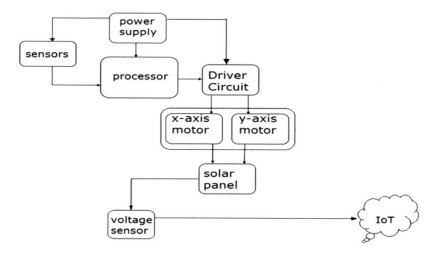

FIGURE 9.1 Overview of dual axis solar tracker.

When there is no signal from the Arduino microcontroller to the driver, then both the common terminals connected to the motor terminals are short-circuited with the NC terminals of the relay. So both the terminals of the motor are in a lower potential (i.e., negative) and the motor remains stationary.

When one of the relays receives the signal from the Arduino microcontroller, the common terminal of the respective relay gets short-circuited with the NO terminal of the relay, thereby attaining a positive charge. The motor rotates in a specific direction which is determined by the direction of current flow between the common terminals. If the other relay gets excited by the signal from the Arduino microcontroller, the motor rotates in the opposite direction with respect to the previously mentioned case.

When both the relays are excited by the signals from the Arduino controller, both the common terminals which were earlier connected to the NC terminals, now get short-circuited with the NO terminals of the relay thereby both the terminals of the motor are in a higher potential (i.e., positive) and the motor does not rotate (Figs. 9.2 and 9.3).

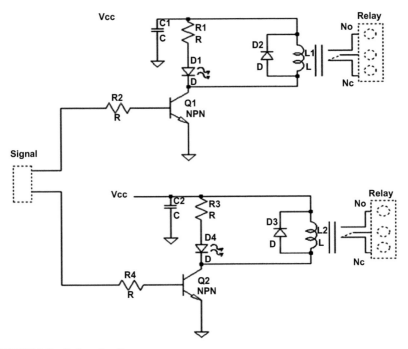

FIGURE 9.2 Relay circuit.

9.3.4 *POWER SUPPLY*

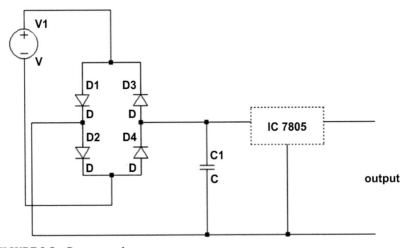

FIGURE 9.3 Power supply.

A 12 V battery is used to power the setup including the Arduino microcontroller, light intensity sensors, and the motors. The 12 V from the battery is converted to a 5 V supply using a 7805 V regulator IC. The 5 V supply is given to the Arduino microcontroller and the light intensity sensors. A 12 V supply is required to run the motors and the relay circuit. The Wi-Fi module is used in the setup for monitoring the solar panel powered by the Arduino microcontroller.

9.3.5 ESP8266 MODULE

The ESP8266 is a Wi-Fi Module. It is a system on a chip with an integrated TCP/IP protocol stack. The protocol stack gives many microcontrollers access to Wi-Fi networks. The device is capable of hosting an application or offloading all Wi-Fi networking functions from another application processor. The module can be easily connected to an Arduino device as it is pre-programmed with an AT command set firmware. The module is a very cost -effective board (Fig. 9.4).

FIGURE 9.4   ESP8266 module.

To transmit the data from Arduino microcontroller to cloud, two pins are interconnected between Arduino and the module. The connections are described in Table 9.1 (Fig. 9.5).

TABLE 9.1   Connection between Arduino and ESP Module.

Arduino	ESP8266 module
TX (digital pin 1)	(Receiver pin for ESP module)
RX (digital pin 0)	(Transmitter pin for ESP module)
GND (ground pin)	GND (ground pin)
VCC (3.3 V)	VCC

TX, transmitter pin; RX, receiver pin.

9.3.6   HARDWARE SETUP

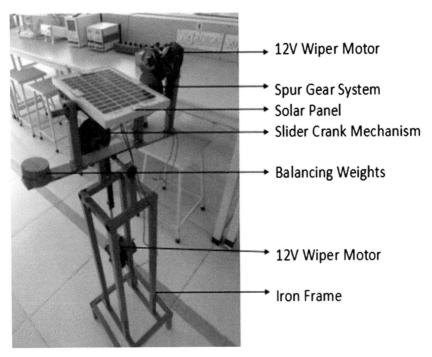

FIGURE 9.5   Hardware model.

9.4   SIMULATION IN SOLIDWORKS

The implementation of the hardware is first simulated in solidworks for proper alignment and provides more insight to the design shown in Figure 9.6.

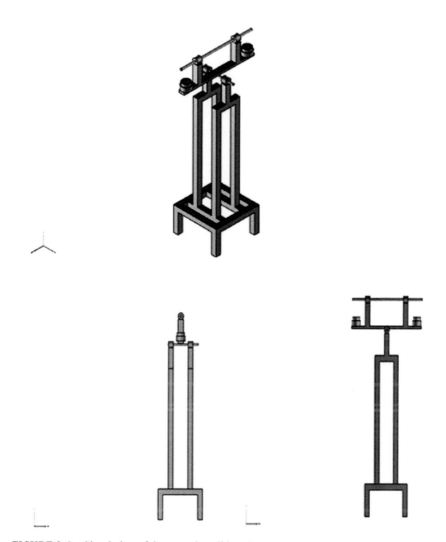

FIGURE 9.6   Simulation of the setup in solidworks.

9.5   MONITORING THROUGH WEBPAGE

Arduino microcontroller calculates the value of voltage obtained each time and transmits the data into the Wi-Fi module which in turn pushes the data to the cloud.[8] These data are displayed on the webpage as shown in Figure 9.7. The time of upload of data can be seen by taking the arrow to the required point. The voltage obtained is stored over time and presented to the user in the form. The experiments show that employing dual-axis tracking system under sunny conditions has resulted in increase of the voltage generated by around 10% in the morning and evening but with no significant change during the afternoon. When, the arrow is taken to a point it shows the voltage value and time at which it is uploaded to the cloud. The URL of the graph is given below "https://thingspeak.com/channels/251825/charts/1?bgcolor=%23ffffff&color=%23d62020&dynamic=true&results=100&type=line"

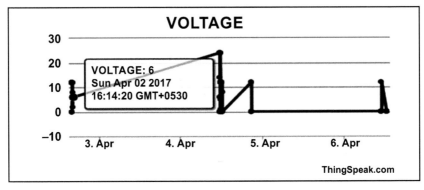

FIGURE 9.7   Webpage.

9.6   ANALYTICAL CALCULATIONS

To make the accuracy of the motor more and increase the resolution there is a need to decrease the speed of the motor. So a spur gear arrangement has been used to decrease the speed of the motor by half. It is only possible to decrease the speed of the motor by half using this arrangement because the size and weight of the gear are constraints because if either of them increases more than this, it will increase the load of the other motor, resulting in malfunction of the system. The signal that is obtained from the voltage sensor is converted using the formula shown below.

Design of gears:

No. of teeth in drive gear ($z1$) = 29
No. of teeth in driven gear ($z2$) = 58
Speed ($N1$) = 60 Rpm

Speed ratio,

$$i = (N1/N2) = (N2/N1) = (60/N2) = 58/29$$

The desired speed of the motor is obtained and is 30 rpm.

The formula for obtaining the voltage measured has to be improved using trial and error method. It is obtained using a pre known value of voltage source and trying to measure it with the voltage sensor.

$$V = ([12 \times res]/1023);$$

The "res" in the above formula is obtained from the voltage sensor which can vary from 0 to 1023.

9.7 EXPERIMENTAL RESULTS

The prototype has been tested under sunny conditions.[9] The voltage sensor can measure voltage approximately and transmit the data for every time. The voltage is being measured to the website. The results are summarized in Table 9.2.

TABLE 9.2 Experimental Results.

Sunny conditions	10:00 AM	2:00 PM	6:00 PM
Without tracking	17.2 V	19.8 V	16.5 V
With tracking	19.5 V	20.4 V	18.3 V
Percentage increase in the efficiency	10.79%	3%	10.90%

9.8 CONCLUSION

The design, implementation, and testing of a dual-axis solar tracking system is presented in the study. The performance of the developed system was studied and compared with the static solar tracking system. The work demonstrates that the dual-axis solar tracking system can assure higher power generation compared to static panel. The experiments show that employing dual-axis tracking system under sunny conditions has resulted in increase of the voltage generated by around 10% in the morning and evening but with no significant change during the afternoon. In the dual-axis solar tracking system, both motors run continuously throughout the day thereby aligning the solar panel with the sun's position. The designed tracking system can also be implemented for solar thermal systems. The proposed design has achieved low power consumption, high accuracy, and low cost.

KEYWORDS

- **renewable energy resources**
- **dual axis**
- **solar tracker**
- **internet of things**

REFERENCES

1. Shen, C.-L.; Tsai, C.-T. Double-linear Approximation Algorithm to Achieve Maximum-Power-Point Tracking for Arrays. *Energies* **2012,** *5* (6), 1982–1997.
2. Mousazadeh, H.; Keyhani, A.; Javadi, A; Mobli, H.; Abrinia, K.; Sharifi, A. A Review of Principle and Sun-tracking for Maximizing Solar Systems Output. *Renew. Sustain. Energy Rev.* **2009,** *13* (8), 1800–1818.
3. Barsoum, N. Fabrication of Dual-axis Solar Tracking Controller Project. *Intell. Control Auto.* **2011,** *2* (2), 57–68.
4. Kelly, N. A.; Gibson, T. L. Increasing the Solar Photovoltaic Energy Capture on Sunny and Cloudy Days. *Solar Energy.* **2011,** *85* (1), 111–125.
5. Benghanem, M. Optimization of Tilt Angle for Solar Panel: Case Study for Madinah, Saudi Arabia. *Appl. Energy.* **2011,** *88* (4), 1427–1433.
6. [Koyuncu, B.; Balasubramanian, K. A Microprocessor Controlled Automatic Sun Tracker. *IEEE Trans. Consum. Electron.* **1991,** *37* (4), 913–917.
7. Divya, M.; Vijaya, R. R. V.; Tharun, S. Smart Dual Axes Solar Tracking. In *Proceedings of the International Conference on Energy Systems and Applications*; ICESA, **2015**; pp 370–374.
8. Peng, Z.; Gongbo, Z.; Zhencai, Z.; Wei, L.; Zhixiong, C. Numerical Study on the Properties of an Active Sun Tracker for Solar Street Light. *Mechatronics* **2013,** *23* (8), 1215–1222. DOI: 10.1016/j.mechatronics.2013.08.007
9. Roth, P.; Georgiev, A.; Boudinov, H. Design and Construction of a System for Sun tracking. *Renew. Energy.* **2004,** *29* (3), 393–402.

CHAPTER 10

Demystifying Next-Generation Cyber-Physical Healthcare Systems

VEERAMUTHU VENKATESH[1], PETHURU RAJ[2], SATHISH A. P. KUMAR[3], SURIYA PRABA T[1*], and R. ANUSHIA DEVI[1]

[1]*School of Computing, SASTRA Deemed University, Thanjavur 613401, India*

[2]*Reliance Jio Cloud Services (JCS), Bangalore 560025, India*

[3]*Department of Computing Sciences, Coastal Carolina University, Conway, SC 29528, USA*

[*]*Corresponding author. E-mail: suriyathiyagarajan03@gmail.com*

ABSTRACT

Digitization of physical, mechanical, and electrical systems gains momentum these days, and it is projected that in the years ahead, there will be trillions of digitized objects/sentient materials. The faster stability and maturity of edge technologies (sensors, actuators, beacons, RFID tags, stickers, barcodes, chips, microcontrollers, smart dust, specks, etc.) lead to the faster and systematic realization of digitized systems, which are typically computational, connected, perceptive, active, reactive, decision-making, etc. Now, these digitized entities are being meticulously integrated with cyber resources (enterprise-scale applications such as enterprise resource planning, supply chain management, employee relationship management, knowledge and content management, etc.), a variety of purpose-specific as well as agnostic micro-services, databases (SQL, NoSQL, In-Memory, time-series, etc.). This unique yet challenging linkage between ground-level digitized elements and cyber-level resources is to result in a series of sophisticated, adaptive, service-oriented, event-driven, insights-filled, people-centric systems, and services. Such kinds of empowered and deeply integrated systems are being touted as cyber-physical systems (CPSs). Having understood the strategic advantages of such extremely connected systems, every industry vertical is consciously embracing this technological trend to be right and relevant to their customers, clients, and consumers. The healthcare industry is displaying a lot of interest in leveraging the widely insisted benefits of CPSs. There are several medical services, applications, and instruments and these can be decisively and deftly enhanced through this unique integration capability. This chapter is specifically being prepared to tell all about the CPS journey and how it is to impact the medical field in bringing forth next-generation and state-of-the-art medical devices and services.

10.1 INTRODUCTION

10.1.1   A BRIEF ON CYBER-PHYSICAL SYSTEMS

Cyber-physical systems (CPSs) can be explained as a technology for maintaining connections between a system's physical resources and computational capabilities.[1] The CPS is made up of several active digitals, analogue, physical, and human parts that are made to work using integrated physics and logic. CPS is well renowned for its potential in several sectors as innovative applications to enhance world economy in fields such as healthcare, construction, manufacturing, supply chain. In the fast-growing world, and huge availability of affordable sensors and networks, industries are competing to take a huge leap in implementing high tech technologies. This has invariably given rise to the collection of an enormous amount of data known as Big Data. In such a given situation, CPS has found use in supporting machine networking and managing large datasets in order to make intelligent machines.

The main features of cyber-physical system include:

- Physical parts with cyber capability
- Wireless and wired networking in large scale
- Extreme and multiple scale networking
- Temporal and spatial complexity
- Recursive and dynamic organization and configuration
- Large scale automation
- High level of security to meet privacy needs

10.1.2   ACHIEVING NEW LEVELS OF ABSTRACTION IN CPS

Embedded and CPSs face several challenges in their development regarding the numerous design practices involved. This becomes difficult, especially because there is no exact language for design practice. To make is easy for developers from all fields to work on the progress of cyber systems for continuous innovation, it is essential to set the software and physical responsibilities right. A procedure manual for the development and deployment of a CPS has been devised. The CPS has two core components: (1) Gathering real-time data from the system and information feedback from cyberspace. (2) Managing, analyzing and computing the collected data to construct the cyberspace. These requirements cannot be implemented as there are not fit for implementation as shown in Figure 10.1.

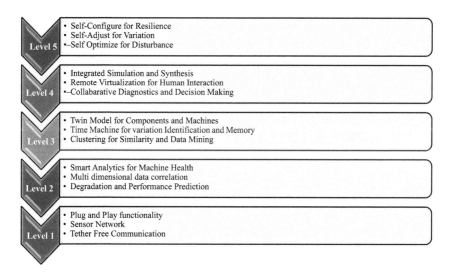

FIGURE 10.1   Different levels of cyber-physical system.

10.1.2.1   DATA ACQUISITION AND CONNECTIVITY LEVEL – 1

Combining and coordinating computational and physical resources are the important features of CPS. The foundation step for developing a CPS includes collecting accurate and trustworthy data from machines. These data can be collected directly from sensor readings or manufacturing systems such as ERP, MES, SCM, and CMM. It is important to consider a variety of data types for a seamless way to manage and transfer the data to the central server.

10.1.2.2   INFORMATION CAPTURING LEVEL – 2

It is important that we can understand the collected data. Several modern methodologies have been introduced for data to information conversion. Several studies are being carried out in the development of algorithms for healthcare management. This aids in calculating health value and an estimate of the remaining useful life, thus bringing self-awareness to machines.

10.1.2.3   CYBER-PHYSICAL LEVEL – 3

The cyber level is the heart of the architecture. It is an information hub into which the collected information from every other connected machine is pushed. When a large amount of data has been gathered, analytics has to be done specifically to give us a better outlook of the status of every individual machine despite the groups of machines. This is advanced, and the analytics gives us a self-comparison of the machine's performance capacity and every other machine connected to it. To foretell the future behavior of the machine, the similarities between the machines' performance and assets are needed.

10.1.2.4   COGNITION FOR CPS LEVEL – 4

Using CPS in the level provides us with a deep knowledge of our system. A good exhibition of the knowledge obtained to skilled users helps us take the right decision, as all the data on the machine performance capability and the comparative information is available, decisions on the priority of tasks to optimize the maintaining process can be made. Info graphics are involved in transferring complete knowledge.

10.1.2.5   CONFIGURATION FOR CPS LEVEL – 5

The cyber to physical space feedback is the configuration level, and this is used to supervise self-configurable and self-adaptive machines. This is a resilience control system (RCS) used to make corrective and preventive decisions.

10.1.3   THE IMPORTANT USE CASES OF CPS

Systems with interactive computational parts and physical elements are CPSs. This provides man with an array of capabilities to control and interact with physical systems. These systems are used not only in industries and healthcare, but also used in automobile mobility.

10.1.3.1   SMART HOME CASE STUDY: BASED ON CYBER-PHYSICAL SYSTEMS

With the developments in the Internet of Things (IoT), CPSs have been used in smart homes.[3] In the near future, it is predicted that CPS will become a major player in the smart home development sector. It is preferable because it is convenient and efficient for regular home residents to live a better life. A need for improved security and privacy challenges such as confidentiality, authenticity, and integrity of the data that have been collected and sensed by the IoT sensors has led to the IoT models entering out homes, always connected to the internet. These challenges prove that machines connected to the internet pose a constant threat of insecurity. To improve security, it is essential to study the possible security risks. A number of challenges need to be faced in developing and implementing the system, such as portability, timing, prediction, and integrity.

Smart home automation is focused on day lighting, heating, ventilation, air conditioning systems, and other environmental services. For safety and security purposes as well as for electricity, saving smart homes have automated solutions as shown in Figure 10.2.

We can categorize home automation systems in two: locally controlled systems and remotely controlled systems. As the name suggests, it is clear that the locally controlled system is nothing but a stationary system and comes with an in-premises controller. All systems can be accessed from within our homes using a device.

As for remotely controlled automation systems, internet connection plays a major role in offering the users complete control of the smart systems from the pc or mobile. This method enables us to integrate with an existing security system using a smart home even when you are physically present. IoT-based smart homes are extremely vulnerable to several diverse security threats. This is unsafe as the user's privacy is at stake if his information gets compromised. To avoid this, we need to take steps to make smart homes more safe and secure. The home system needs to be carefully studied to assess all possibilities of security risks.

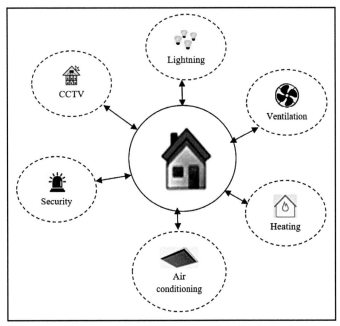

FIGURE 10.2   Smart home environment based on cyber-physical system.

10.1.3.2   *SECURITY FOR HEALTHCARE CASE STUDY: BASED ON CYBER-PHYSICAL SYSTEMS*

Utilizing real-time monitoring systems in healthcare has enhanced personal care in more ways than one. But this will include the usage of several algorithms and CPSs. A study shows that using wireless technology in healthcare systems can increase the efficiency of hospitals.[4] A major example would be a system that generates reminders to patients who need to stick to their medication charts for medical conditions like Diabetes, Asthma, and HIV thus reducing the risk of hypoglycemic episodes, asthma attacks, or blood count deterioration. Other options to track daily activities, detect fall and movements, track patient location, period medicine intake is some of the features available in Ambient Assisted Living (AAL). Industry 4.0 has introduced eHealthcare systems composed of multiple components based on CPSs. There is integration between patients and virtual components given in Figure 10.3. The data are collected from the patient's body using biosensors and other equipment like ECG and EEG. The data collected is then processed by the CPSs using artificial organs such as cardiac pacemakers, insulin pumps, and endoprosthesis as knee and hip implants.

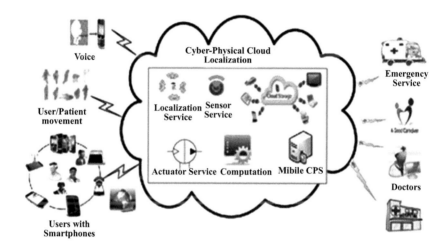

FIGURE 10.3   eHealthcare monitoring system using cyber-physical system.

10.2   COMMON THREATS TO HEALTHCARE

It is very important to maintain a high level of security for health records to monitor the storage and retrieval of data. This is because any kind of violation in confidentiality and privacy or integrity of patient information can lead to sever significances to the patient's life. This is a very serious issue as any failure in obtaining accurate medical data by the WBAN can affect the patient's treatment leading to lethal situations.

In emergencies, if the eHealthcare system is come under the Denial of Service, this is very risky for the patient's life as we might miss out vital signs. An intruder can meddle with the data generating malicious activities in the network and can cause damage to the patient's monitoring.

Service-oriented architecture is used in distributed eHealthcare systems. Web services are being provided not only to patients, doctors, and nurses, but also to medical instruments monitoring patients. The main components of the system comprise of a PDA, Web Server, doctor PDA/computer, patient PDA, and Bluetooth for sensor communication. This comes with extended security features such as user authentication and period information logging. But there is a problem in logging the data for offline uses along with no aid in emergency scenarios as in HIPAA. There are no integrity checks, no handling of availability issues, and no pseudonymization of the patient data.

10.3 SECURING CYBER-PHYSICAL HEALTHCARE NETWORKS

Our eHealthcare systems have developed from WBANs to CPS because of the research growth in wireless and sensor technologies. CPS leverages sensing, processing, and networking technologies in hosting expensive personalized healthcare systems. These systems make intelligent decisions based on massive patient data. Not only the security and privacy of the patient medical records are important, but also the confidentiality of the patient details can be very important in terms of cyber-physical healthcare. A very tight security system needs to be in place. A mutual authentication has to be established between the wearables and the system. The proper security system needs to be in place to secure the core components and connected networks. A result of such good security, this scheme is taken into consideration when making plans for security schemes. Also, kindness should be given to taking advantage of NFV and SDN in deploying these schemes.

10.4 VEHICULAR CYBER-PHYSICAL SYSTEMS CASE STUDY

The rising environmental problem is traffic and congestion problems. With a large number of vehicles being manufactured each year, the transportation sector has grown significantly leading to an increase in the number of accidents and fatalities.[5] This has also had adverse effects on the environment and the worldwide economy. To address these problems including reducing in traffic jam, decrease in congestion, putting a check on fuel consumption, and to save the time spent in traffic jams, smart transportation using CPSs have been deployed. The smart transportation system has played a major part in the design and development of intelligent transportation systems. The major developments in embedded systems, sensor networks, and wireless communications help us bridge together physical components with the cyber world. In CPS, there is an exchange of information between vehicles (V2V) via vehicular *ad hoc* network (VANET) and vehicle-to-roadside (V2R) communications given in Figure 10.4. For wireless communication, IEEE has an 802.11p standard for Dedicated Short-Range Communication (DSRC) for Wireless Access for Vehicular Environment (WAVE).

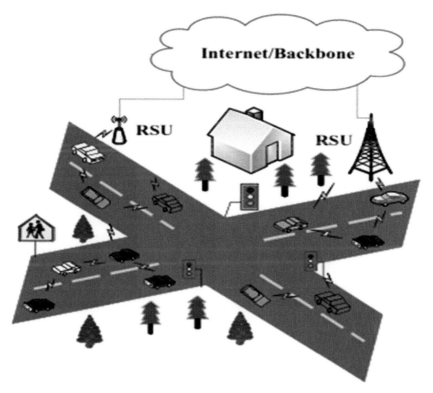

FIGURE 10.4 CPS based vehicle-to-roadside communications for smart transportation.

10.5   COMMUNICATION FOR SMART TRANSPORTATION CPS

The three essential parts of any intelligent system are computing, communication, and networking. There is an exchange of information between vehicles using *ad hoc* based peer to peer networks in road-side infrastructure.

Also, VANET vehicles can use public or hybrid or private cloud for computing, i.e., illustrated in Figure 10.4. The information collected is processed by forming a VANET cloud of vehicles for collaborative processing or hybrid of VANET and public cloud.

10.6   A LOOK ON THE VARIOUS LIMITATIONS OF CURRENT HEALTHCARE SYSTEMS AND APPLICATIONS

The beginning of a new generation of WSN has been marked by the evolution of Wireless Body Area Network (WBANs).[6] It has been primarily dedicated to the healthcare industry and aims at keeping a check on a patient's vital parameters by monitoring them continuously while also proving them with the ability to move around freely. This promotes the healthcare facility to another level. The internal sensors and body surfaces are crucial for advanced healthcare delivery.

Nonetheless, there are a number of technical challenges when it comes to meeting the wireless sensor network's complete potential in this industry. These technical challenges are not just limited to the resource constraints; rather, they extend to memory and processing constraints, limited network capacity, as well as scarce energy reserves. This domain also demands required system reliability, the service quality, and also needs a mechanism to protect information for privacy reasons, unlike the other domains or applications.

10.7   REQUIREMENTS FOR WIRELESS MEDICAL SENSORS

Wearable and implantable sensors are used to sense the biological information from the patient's body and transmit it across a minimal distance. The sensors transmit the acquired data of the human body to a control device that is also worn by the patient or is kept at an accessible distance from them. This information from the control devices is then in turn transmitted to a wireless body-area network for diagnosis in order to perform the required therapeutic action if needed by involving other long-range wireless networks. There are a number of requirements that a network for healthcare needs to be satisfied to serve the purpose for which it was developed.

10.8   UNOBTRUSIVENESS

One of the most important requirements for a medical sensor is that they should be of light weight and a considerably smaller size to allow non-invasive and unobstructed monitoring. These characteristics are directly related to the sensor's battery capacity, and its size as the capacity is directly related to its size. Advancement in technology in terms of the development of on-chip system design, microelectronics, and wireless communication with minimal energy consumption has led to the birth of high capacity batteries. Printed batteries are highly flexible as they can be used for a variety of applications like transdermal delivery of the drug, sensing the temperature of the body as well as RFID tracking.

10.9   POWER EFFICIENCY AND ENERGY SCAVENGING

One other challenge in the deployment of WWHMS is enabling a non-attractive or obvious operation for a considerable time. Energy consumption in the wireless systems are due to the following operations: transmission of a data packet, receiving a data packet, observing radio signals, sampling the sensor, reading data from the ADC and flash memory and performing a write or an erase operation on the flash memory. One proposed method addresses these issues in the MAC layer, where it coordinates node access with the wireless medium being shared. A MAC that is energy-efficient must support a wide range of applications as well as data types, such as periodic, continuous, nonperiodic, and burst while ensuring a considerable level of quality in service. The WWHMS utilizes a star topology with the central node as the coordinator, which is capable of ensuring sufficient computing resources in the latter as observed in their search prototypes. The coordinating node is given the power to activate or deactivate a slave node to prevent asymmetric uplink and traffic in the downlink and also for effective power management. When a slave node is activated or is awake, it senses the patient's body and monitors the vitals and generates data. It can process the acquired information using the small computing component present in it and then transmits it to the coordinating of the central node by making use of the radio signals and is also capable of receiving data from another node. Once it has performed its task, it is deactivated and put to sleep again. Several other techniques apart from this like making use of low power consumption transceivers, power-scavenging techniques also promises a greater power saving when combined with the existing ones.

10.10   ROBUST COMMUNICATIONS AND HIGH THROUGHPUT NETWORKS

When the signal received by a node is degraded due to environmental obstructions such as temporal physical obstructions deep fading occurs. WWHMSs especially are likely to face problems of these kinds. There are also factors that cause the degradation of the signal apart from environmental obstructions like interference of the signal; by another network operating on the same radio spectrum. All these might cause discontinuous transmission of data and affect the network's performance and reduce its efficiency. Combating this problem is important in applications like the healthcare, which includes unobstructed monitoring of the patient's body because it might be a life or death chance we are dealing with. In order to save the power of the primary transceiver, sometimes a secondary transceiver is used by deactivating the primary transceiver. Issues related to the communication between the in- and off-body links can be reduced by making use of diverse channels. This can be a more challenging problem for the radio signals can be weakened around the patient's body at various degrees depending on the body's orientation and size and movement. A large amount of signal passing through the body can be hazardous, and thus we might also want to make sure a less amount of radio signals passes through the body so that the treatment can be fulfilled without any further surgical adjustments. Some treatments that make use of a camera to monitor the patient's body necessitate a good resolution camera to be functional inside the human body for higher accuracy.

10.10.1   GREEN COMMUNICATION

There has always been a need to ensure a network requires minimum power to remain operational, and this need has arisen a number of issues in terms of providing an unobstructed flow of power through the nodes. Hence, we have adopted a number of communication protocols to save energy and also communicate efficiently and reliably. As the CPS domain evolves, their applications have become wide spread, and hence there is a greater pressure to ensure the use of energy harvesting mechanisms to control the environment they have been deployed in. The safety and efficiency, particularly in terms of energy, should be high for CPSs and M2M as they directly interact with the environment. Their safety and performance are of immense priority. Researches that focus on M2M and integrating them with the environment are essential for the enhancements in these domains. Despite providing guidance for adopting energy efficient mechanisms, there are also researches going on to discover new power sources that are produced by causing minimal harm to the environment aids in green IoT. These sources can be solar, thermal, wind, or any energy produced from natural sources.

10.10.2   INTEROPERABILITY

As far as healthcare is concerned interoperability describes the limit up to which the data are being interpreted and analyzed and present it to the users in a meaningful and understandable way. This includes the exchange of information between hospitals, pharmacies, clinics, as well as between patients. Interoperability aims at providing mobile information. Information once fed into a system should be understandable and exchangeable between any two bodies. In short, interoperability makes sure that any type of information is easily exchanged.

10.10.3   RELIABLE COMMUNICATION

In applications like the healthcare reliability of the link between the nodes should be really high. The issues that arise in this area are because the sensor's sampling rate differs. Typically, in a place electrocardiogram (ECG) is being used, we might not need to transmit the raw data; instead, we can extract only the information that is useful and transmit it. This also saves the battery life as now there is not much demand on the communication channel. Hence, one might have to consider the tradeoffs between the computation and the transmission for efficient system design.

10.10.4 SECURITY

Like in any network, there should always be some mechanisms included ensuring integrity and security. Privacy of the data should be maintained. The software should be designed such that only authenticated body can access private information. Including measures to ensure data security is essential like any other element of a network. There needs to be better coordination between the hardware and software to fulfill the above mentioned.

10.10.5 PRIOR ART

CPSs employed in the medical field are nothing but an integration of several medical devices used for various purposes.[7] These networks are being used in hospitals to improve the quality of treatment and also to make a few things easier for both doctors and patients. Although their advantages are numerous, they do face several challenges like security or privacy issues, inoperability. This chapter aims at improving the quality and well the security of the networked Medical Devices (MD). This network assists the medical specialist in monitoring the vitals of the patients while they are on the move and also provides this information in a secure manner. Secure social networking combined with Wireless Sensor Networks (WSNs) is being discussed. Once the environment and the platform for the establishment of the network have been established, we move our focus toward CPS-based services in healthcare. This chapter particularly highlights the changes that can be brought about in the healthcare domain when CPS is integrated with it. It also highlights the kind of infrastructure needed to achieve this real time.

10.11 CYBER-PHYSICAL SYSTEMS AND MEDICAL CPS STRUCTURE

A huge economic loss might be incurred if the CPS malfunctions and immediately has an impact on the operation of the system. The physical infrastructure requirements have to be met long with the networking models. The network becomes more sophisticated and prevalent, customized and reliable. The application is not just limited to the medical domain then; it can also be applied to smart grids and robotics. Improvements in CPS will enhance the capability, adaptability, usability, security and safety, and resiliency, resulting in a completely upgraded system different from the simplest of the embedded systems.

In the case of online treatments, interaction between the patients and the doctors takes place through a global platform. Hence, there is a need to secure the patient's report, and details that are being passed through the platform for the information is highly confidential. Thus, one must ensure that there is no unauthorized flow of data and provide inclusion of necessary features in it.

Researchers all around the globe have been inspired by a different kind of interpretation of the CPSs and hence, a concrete pattern and a clear definition are still evolving as research progresses. In general, CPS and Medical CPS (MCPS) are considered to be physical, engineered systems whose processes are being monitored, controlled, coordinated and securely integrated by a communication and computing core at all levels and scales. CPS these days are working in a social environment where there is a higher degree of human inclusion and interaction. Henceforth, any application that includes CPS and MCPS in them will take humans into consideration during the development phase itself. Thus, drivers can provide data by sharing information about traffic blockage in their area; patients can have their vitals monitored by doctors even when they are on the move and survivors can share information about the cause of the natural disaster. MCPS implementing by passing medical data to CPS as shown in Figure 10.5.

10.11.1 MEDICAL CYBER-PHYSICAL SYSTEM

Social media refers to the facilities available on the web that allows easy communication and connections worldwide between individuals. It contributes to the MCPS in a way it helps patients share their reports online and interact with their physicians to get their immediate medicines prescribed and have quick check-ups. It has been proven to be a venue for interaction between the doctor and the patient. This has led to an increased role of social media in the medical environment. On the other hand, SPCS has also become well integrated with the medical sector as MCPS has integrated software and networking connectivity.

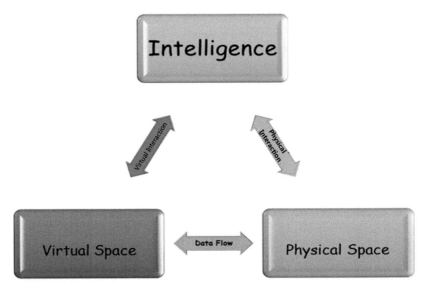

FIGURE 10.5   Medical CPS system (MCPS).

10.11.2   *CPS REALIZATION TECHNOLOGIES AND TOOLS*

The existing technology developed in the early stages are not sufficient to accommodate the latest developments in technology.[8] To keep up with the growth in technology and to keep this standard followed up to the mark is a task. In order to meet the rapid change in the situations, the industries are forced to constantly modify and customize their products to make them more responsive. Flexible systems are capable of handling agile change at a reasonable cost with real-time reactivity. Aiming at developing and implementing tools and methods for achieving flexible and scalable interoperable network enabled collaboration between decentralized and distributed embedded devices and systems; the "Collaborative automation" paradigm is used. This trend is accompanied by technological growth marked by penetration capabilities. This includes data processing that slowly transforms the shop floor into a networked ecosystem that comprises of embedded devices and systems. In businesses, customers and business partners have an interaction with the physical system in pursuit of a well-defined system. The key innovation in the cyber-physical industrial domain is termed as industrial CPS. A number of communities and forums are researching on designing, implementing, and assessing ways to realize CPS. Industrial automation is supported by developing the technology needed such as smart buildings, smart transportation, smart healthcare, and many more.

10.11.3   CPS ENGINEERING APPROACHES AND TOOLS

A greater part of the contemporary tools used are vendor precise and abutment closed control environments. With respect to limited agility in delivering sturdy operational systems, CPS offers good point solution functionality. The technical working group called the WG2 is researching in already available approaches and methods using Industry 4.0 in order to identify new approaches. But this does not entirely address the aspects of Industry 4.0. Aimed at end-to-end engineering, Automation ML, ProSTEP iViP, and eCl@ss are advised of utmost interest. However for intelligent transmission, the OPC-UA-based IEC 62541 and the data centric layer, field device integration (FDI) are used as alliance technology.

That keeps up the congeniality with the existing conventional standards, and to start off the definition of new ones CPS has to be used. Noteworthy standards as of an engineering viewpoint include IEC62264 enterprise control system alliance, ISA Draft 88/95 technical description, using ISA88 and ISA95 together, IEC 62890 lifecycle and cost stream, IEC 62264/IEC 61512 grading levels. In order to dynamically expand the role of each player in a CPS automation process, open platforms and tools are used by deranging old business models. The CPS standards are very crucial in accelerating the implementation of CPS engineering tools' development.

10.11.4   NEXT-GENERATION CYBER-PHYSICAL SYSTEMS

The CPS domain is growing at a rapid pace these days with the continuous advancements in the aspect of device connectivity. The device ecosystem is growing steadily and hence, we are experiencing slim and sleek, handy and trendy and multi-faceted devices capable of doing many things for people. Different and distributed devices are connected and integrated with cloud-hosted applications and databases. Devices, through their connectivity and the seamless integration with faraway applications, are becoming smarter in their actions and reactions. Device-to-device (D2D) and device-to-cloud (D2C) integrations become popular and pervasive. Physical devices at the ground level are all set to become cognitive in their operations, offerings, and outputs. Further on, all kinds of physical, mechanical, and electrical systems are also attached with edge devices to make them digitized. Thus, our everyday environments are all set to be fully digitized with the smart application of a variety of digitization and digitalization technologies.

On the other hand, the new phenomenon of cloud-native computing is gaining a lot of momentum with the faster maturity and stability of multiple contributory and assistive technologies such as virtualization, containerization, microservices architecture, orchestration platforms, analytics platforms, machine learning toolkits, etc. Thus, next-generation CPSs are supremely intelligent. The connectivity and the cognition traits make every physical system to be smart in their contributions. CPSs are therefore designed and destined to be adaptive, adjustive, accommodative, assistive, and adroit. The distinct innovations and disruptions in the information and communication technologies (ICT) landscape are resulting in advanced CPSs across industry verticals.

The understanding of CPS takes tangible shape in several sectors, there is a perceptible change being experienced by businesses and people. CPS also contributes exceedingly in connecting and empowering more and more complex social systems such as the environment, energy, medical care, safety, economy, and transportation. A number of manual activities are getting automated, accelerated, and augmented. As enterprises embrace the much-discussed digital transformation initiatives, the role and responsibility of CPS are going to be very crucial. CPS is one of the most popular technological paradigms intrinsically capable of creating and sustaining a number of digitally enabled transformations. A number of improvisations in the technology space are greatly contributing for the runaway success of the CPS paradigm.

The realization of the high-speed processors suited for processing massive volumes of multi-structured data along with the widespread use of IoT tools feeds the faster maturity and stability of the CPS concept. The emergence of clouds as the one-stop IT solution for all kinds of business requirements is another key driver for the unprecedented adoption of CPSs. A variety of sensors and actuators are involved to the control target (person, car, manufacturing apparatus, etc.) in order to collect raw data about various status updates, breakthroughs, threshold break-ins, transitions, and other exemplary phenomena. The gleaned data then are being subjected to a variety of deeper and decisive investigations to extricate useful and usable insights, which can be looped back to empower our everyday systems, appliances, devices, equipment, instruments, wares, utilities, consumer electronics, drones, robots, etc. With the ready availability of big, fast, and streaming data analytics platforms, every bit of data being gathered data is being turned into information and knowledge. This is the widely accepted and adopted method to bring forth smarter and sophisticated systems in order to assist us in our everyday tasks. Premium, people-centric and path-breaking services can be conceived and concretized toward digitally transformed cities, factories, grids, hospitals, hotels, homes, airports, etc.

The next-generation CPSs are going to be innately intelligent. Based on the literature review, the most relevant core elements are identified and summarized as follows:

Connectedness and Cognition – This is the basic requirement to design, develop, and deploy CPSs, which are the primary pillar for the vision of Industry 4.0. The other related term which is widely used known as smart manufacturing. Once physical systems are connected locally and remotely with cyber applications, the various interactions, collaborations, corroborations, and correlations result in a huge amount of data. Thus, data analytics and artificial intelligence (AI) algorithms play a very vital role in realizing intelligent CPSs. Thus, CPS data have to be consciously and carefully collected to untangle any hidden patterns, associations, insights, etc. Business processes get optimized; systems become smarter in their decision-making, deals, and deeds; company revenue is to go up; customer delight is to be achieved; productivity can be increased substantially; and new possibilities and opportunities can be realized, etc.

The horizontal consolidation typically links numerous stages of manufacturing and business planning methods and their corresponding IT systems. This allows the interchange of information which can moreover take place within a single company or interlink diverse companies leading to networks. On the other hand, the vertical integration defines the integration of diverse IT systems which are used at discrete hierarchical levels, for example, production control or business planning levels. Together, we get an end-to-end IT solution.

Smart Machines and Smart Products – IBM prescribes three things. Instrumentation, interconnection, and intelligence are the main prerequisites for establishing and sustaining smarter environments. We need to have smart machines and products in order to realize CPSs in plenty in our personal as well as professional environments. Machines ought to be empowered to find, bind, and communicate with one another. Further on, they need to get hooked to web and cloud-based applications and data sources. Physical, mechanical, and electrical systems have to be attached with additional components for enabling local as well as remote communication. There are specialized communication modules and gateways for enabling ordinary systems to be digitized and to join in the mainstream computing. There are sensors and actuators to be implanted on our machines and products to be adequately empowered to be categorized as a CPS. CPSs can take data from their environments and interact with the environment. They can ceaselessly check whether the process parameters are correct. If not, they can identify the deviations and deficiencies and correct them in an automated manner.

Decentralization – This is an important factor for the greater success of the CPS idea. Decentralization refers to the control of the scheduling in how products know their operations. Also, the sequence of the process is very important. The idea is that autonomous and cooperative products intelligently accomplish their assigned works. They interact with one another, find their unique capabilities, and accordingly complete the process-centric jobs. This decentralization empowers shoo-level decision-making and work completion. There is no centralized module to monitor, measure, and manage all participating products. The negotiation between work pieces and machines takes place based on the requirements of the work piece and the capabilities of the production entities. Thus, decentralized planning and execution are very essential for achieving the intended success of the CPS paradigm.

Cybersecurity – When any tangible and concrete article gets connected, there is a great possibility of increasing security threats. Thus, the cybersecurity discipline becomes an important component of CPSs. Researchers and practitioners have to intently and intensively focus on unearthing competent security management mechanisms so that CPSs cannot be misused to get any kind of corporate, confidential, and customer data. With the widespread availability of the Internet, which is the most affordable, open, and public communication infrastructure, the cyber-attacks on CPSs are going to be prevalent.

Thus, device connectivity, individual empowerment, decentralized cooperation, and work execution are being touted as the supreme hallmarks of next-generation CPSs. The convergence of AI and CPS is to result in pioneering CPSs in the years to unfurl. The various characteristics of physical, cyber, and CPSs are portrayed and pictorially represented in the diagram below.

A Systems Context

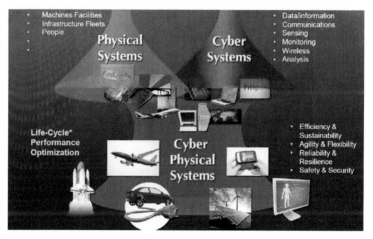

FIGURE 10.6   Seamless interactions between physical and cyber systems.

Precisely speaking, the CPS technology is destined to bring forth radical changes through seamless and spontaneous interactions between physical and cyber systems in Figure 10.6. This linkage is not only capable of producing hitherto unheard use cases and applications for industries, but also for commoners. There are many sectors getting ready to be positively impacted by the application of the CPS technology. New business applications, models, capabilities, and value will be ultimately realized through the continued journey of this strategically sound CPS paradigm. It is anticipated that the connected world will get innumerable benefits. And CPS is to play a better role in determining up the dreams of the human society in the years ahead. Diverse industrial sectors and application fields such as computerized production technology, automotive and mechanical engineering, logistics, energy, and telemedicine are getting immensely benefited out of this unique technological paradigm. In addition, CPS is expected to play a key role in improving and protecting the environment by reducing carbon dioxide emission to our fragile environment. Less electricity consumption in order to dissipate less heat to our environment is one of the unannounced advantages of the drastic advancements happening in the CPS arena.

As we all know, the fourth industrial revolution is characterized by extreme connectivity and superintelligence. The series of innovations in the IoT and AI spaces along with data analytics solutions have brought in a sea change for industrial establishments. The optimized and organized IT infrastructures represented through the cloud technology along with the sharp improvements in the robotics discipline are to bring in many benefits to manufacturing floors. The miniaturization movement kickstarted by the promise of nanotechnology, microelectronics, system on chip (SoC), infinitesimal sensors, etc. handsomely contribute for the cause of next-generation industrial environments. Technologies get cognitively chosen and converged together to achieve more with less. Thus, the domain of CPS is all set to be nourished in a big way in order to facilitate and fulfill the varied goals of smart manufacturing. And which consume an effect on the society extensive and faster than endlessly compared to obtainable industrial revolutions. AI and IoT, big data, cloud system will be converted into the existing industry and provision and be united with high-tech such as robotics, nanotechnology, and biotechnology. The products and service will be linked to a network and the whole thing will be intellectualized.

As the Internet has revolutionized interpersonal communication and interaction, CPS is expected to bring about radical changes in the interaction between the physical and virtual worlds. This is an interesting phenomenon because multiple components join and collaborate with one another in order to conceive and implement novel services to human beings. Also, with the combination of physical and IT things, the accuracy of the IT-enabled decision-making capability is going to sharply rise. Networked-embedded systems in consonance with CPSs are tremendously crucial for the world. These empowered systems are going to the most indispensable infrastructure for supporting and sustaining a variety of critical requirements for our countries and for their safety and security. There will be a number of positive repercussions for the society. Most of the defense equipment are started to produce with CPS. The critical infrastructure, such as the electric grid, pipelines (water and gas), and transportation systems are being strictly implemented and deployed as CPS. The shift to digital control is to lead a series of innovations in our industrial setup.

10.12 THE CPS APPLICATION ARCHITECTURE

Multiple domains are keenly embracing the CPS domain to get benefited immensely out of all the delectable advancements. Here is macrolevel CPS architecture for producing intelligent applications across industry verticals is given in Figure 10.7.

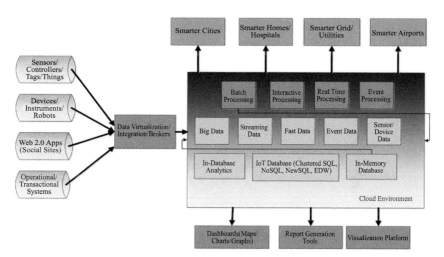

FIGURE 10.7 CPS architecture for intelligent application.

10.13 SECURING CYBER-PHYSICAL SYSTEMS

As enunciated above, the extreme connectivity is liable for remote attacks. Our precarious infrastructure and CPSs are constantly vulnerable to physical attack. But for such massive systems, physical circulation and the energy necessary to extinguish or undermine them is a kind of fortification. Though, attacks through cyber infrastructure can be inexpensive, and networks can deliver those attacks throughout the world very rapidly. Cyberattacks can take the control of CPSs. This can lead to catastrophic consequences in military and civil environments. With the interconnectedness of CPS, such as automobiles, weapon systems, and critical infrastructure, any online attack is going to be a serious issue. As far as the safety and security is concerned, we all know IT infrastructures and assets are being given extra thrust these days as IT has turned out to be a critical component for human living. We have competent security solutions for our communication networks. Data at rest, transit and persistence are secured through a host of path-breaking security mechanisms.

Besides, there are firewalls, intrusion detection and prevention systems (IDS/IPS) to secure our network infrastructures. These come handy in identifying the incidence of intrusions. This approach has a number of intrinsic limitations. Existing IDS can only rely on target system's basic monitoring capabilities. But we need more. There is a need to do feature construction. Anomaly or outlier detection is an important factor for tomorrow's IDS. CPSs are advanced when compared to current computer systems. These advancements can be misused grossly by cyber-attackers. Thus, cyber terrorism is bound to grow in the days with deeper and decisive connected systems. With more functionality being attached in the next-generation CPSs, the security threats and vulnerabilities are to go up. Thus, we need state-of-the-art infrastructures and breakthrough technologies in order to ensure the impenetrable security for CPSs, which are increasingly being used in critical junctions.

10.14   THE IOT INTEGRATION STACK AND THE BIG PICTURE

As discussed before, several promising and potential technologies are being clubbed together to bring a unified technology framework to visualize and realize bigger things. The IoT is one of the key technology paradigms for our future. The integration stack is given below. This architectural illustration clearly articulates the current and emerging data-generating systems. Further on, it vividly conveys various data processing, brokering, and stocking modules. This stack tells how the physical world gets integrated with the virtual world in order to derive intelligent applications.

There will be three dominant spaces: embedded, cloud, and enterprise spaces in Figure 8(a) and (b). All the modules of these spaces are going to be interlinked in order to achieve sophisticated and smarter systems, applications, and services.

The IoT Integration Stack

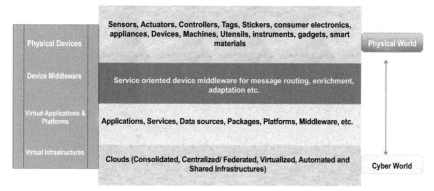

FIGURE 10.8(a) CPS three dominant spaces: embedded, cloud, and enterprise spaces.

The Big Picture

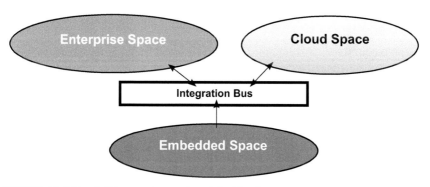

FIGURE 10.8(b) IoT integration stack and the big picture.

10.15 CONCLUSION

The world is heading toward the era of cloud-hosted, event-driven, service-oriented, insights-filled, and knowledge-based services, which are being touted as adaptive, agile, versatile, and resilient. Every common thing in our midst hereafter will be looked up as one supplying intelligent services consistently. The seamless integration between the physical and the cyber worlds is going to be a game-changer not only for businesses, but also for every human being. This integration pours out a lot of data, which has to be subjected to a variety of investigations to extricate actionable insights. The discovered knowledge can be then disseminated to appropriate systems and services to exhibit a kind of adroitness with all the alacrity and clarity. Most of our everyday activities will be technologically auto-mated. Digital articles, connected devices, software services, and cloud infrastructures collectively are bound to contribute exceedingly for our well-being. Precisely speaking, CPSs will be an integral part of our live henceforth. The domain of CPS is all set to see further profound growth toward envisioning and experiencing greater things in the days to unfurl.

KEYWORDS

- **CPS**
- **RFID**
- **sensor**
- **SQL**
- **security**
- **medical CPS**

REFERENCES

1. Liu, Y.; Peng, Y.; Wang, B.; Yao, S.; Liu, Z. Review on Cyber-Physical Systems. In *IEEE/CAA J. Autom. Sinica* **2017,** *4* (1), 27–40.
2. Sha, L.; Gopalakrishnan, S.; Liu, X.; Wang, Q. Cyber-Physical Systems: A New Frontier. *2008 IEEE International Conference on Sensor Networks, Ubiquitous, and Trustworthy Computing (sutc 2008)*; Taichung, 2008; pp 1–9.
3. Seiger, R.; Keller, C.; Niebling, F.; Schlegel, T. Modelling Complex and Flexible Processes for Smart Cyber-Physical Environments. *J. Comput. Sci.* **2015,** *10*, 137–148.
4. Zhang, Y.; Qiu, M.; Tsai, C.; Hassan, M. M.; Alamri, A. Health-CPS: Healthcare Cyber-Physical System Assisted by Cloud and Big Data. In *IEEE Syst. J.* **2017,** *11* (1), 88–95.
5. Wang, S.; Lei, T.; Zhang, L.; Hsu, C.; Yang, F. Offloading Mobile Data Traffic for QoS-aware Service Provision in Vehicular Cyber-Physical Systems. *Future Gener. Comp. Syst.* **2016,** *61*, 118–127.

6. Stankovic, J. A. Research Directions for Cyber Physical Systems in Wireless and Mobile Healthcare. *ACM Trans. Cyber-Phys. Sy*st **2017,** *1* (1), 1:1–1:12.
7. Lee, I. et al. Challenges and Research Directions in Medical Cyber-Physical Systems. *Proc. IEEE* **2012,** *100* (1), 75–90.
8. Harrison, R.; Vera, D.; Ahmad, B. Engineering Methods and Tools for Cyber–Physical Automation Systems. *Proc. IEEE* **2016,** *104* (5), 973–985.

CHAPTER 11

A Novel Cyber-Security Approach for Nodal Authentication in IoT Using Dual VPN Tunneling

N. M. SARAVANA KUMAR[1*], S. BALAMURUGAN[2], K. HARI PRASATH[3], and A. KAVINYA[3]

[1]Department of Computer Science & Engineering, Vivekanandha College of Engineering for Women, Namakkal, Tamil Nadu, India

[2]Founder & Chairman - Albert Einstein Engineering and Research Labs (AEER Labs); Vice Chairman-Renewable Energy Society of India (RESI), India

[3]Department of Information Technology, Vivekanandha College of Engineering for Women, Namakkal, Tamil Nadu, India

*Corresponding author. E-mail: saravanakumaar2008@gmail.com

ABSTRACT

In the history of the planet Earth, the human race is up against an unimaginable counterpart of the present-day civilization that constitutes more associated devices outnumbering the number of people by 2025. It has been projected that of about 25 billion smart things were already been into existent to the estimated 8.1 billion humans. With an advent of Internet in late nineteenth century, the arduous and time-consuming tasks were addressed with negligible time complexity and the strength of the technology is that it enables a seamless access to almost everyone who uses internet in the globe. With Internet, Connection is possible among the people who are sitting at the two different poles of the world. One of the biggest security concern of the deployment and the commercialization of Internet of Things (IoT) is its own underlying data handling policies and its related compromises in the deployment architectures. Since anyone can access the sensors to spoof and thereby eaves dropping the entire IoT model is also possible and the site would be at high risk. This chapter attempts to propose a novel data handling architecture and policies so as to ensure the security of the model at its highest level. With the advent of Internet technologies and its applications, the increase in the need of data storage, analysis, and accessing becomes inevitable. The idea is to deploy a layered access approach to allocate resources using double encryption using VPN to ensure data security over traffic. In this chapter, the two different methods of providing security to the data communicated over IoT channels have been compared with respect to their respective average response time to the various traffic levels.

11.1   INTRODUCTION

Internet of Things (IoT) can be demarcated as an environment consti-tuting various kinds of digital entities representing physical objects that can be inter-connected and mutually accessible among themselves over a dedicated network. The intent of the "IoT" technology leap is to push the limits of the power of the internet beyond computers and smartphones to real-time things and environment. In the current technological era, the things that were participated in an IoT system ranges from a person wearing a smartwatch or having a smartphone or a vehicle with built-in-sensors or electrical appliances or even a traffic light. In this environ-ment, every "thing" was individually assigned an IP address to enable them as a unique entity to identify seamlessly by other "things" and communicate with other things. This makes the IoT as an autonomous smart network that acquires data over a network without any kind of human intervention.

An IoT node is an entity that is capable of aggregating data from the other things of the same network and is then transmitted to a centralized end-user/autonomous node that performs analysis over the data to extract the required information and communicate the corresponding actions/ commands to the applications/devices through the same aggregator (Fig. 11.1). In reinforcement and utility-based expert systems, the under-lying technology embedded in the "things" enables the interaction with internal states and the external environment as a data-centric and informa-tive so that decisions were taken to act upon the environment appropriately.

This perceived information plays a very pivotal role in detecting as many real-time complex and time-consuming patterns, make sensible recommendations from a huge data or scientific data using appropriate statistical algorithms and can possibly be predicted the passive problems before they befall. With the insight provided by advanced analytical algorithms combines with the ability of the smart objects to effectively communicate, the novel systems can be intuited to automate certain repetitive, mundane, laborious, or hazardous tasks.

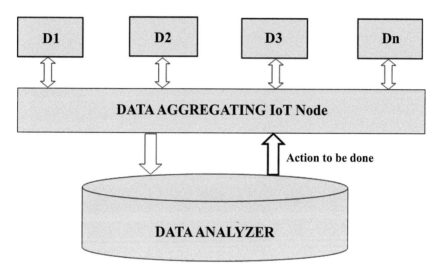

FIGURE 11.1 Data driven IoT stack.

In January 2019, one of the biggest hacking attacks has been taken place and the team released a huge pile of authentication credentials that includes 21 million passwords and 770 million email addresses of almost 87 gigs of data. It is done by creating a virtual traffic using flooding of about 12,000 file requests to the cloud hosting service called MEGA. Till then, it was considered as the largest authentication compromise attack that published tons of user credentials. Soon after a few weeks, by February 2019, a further six collections of about more than 500 GB of log-in credentials data were published online from various sources that shook the world and clearly proclaim that still the network and firewalls have to go long way. This breach also aims IoT ecosystems to this time that evident the vulnerability of the technology.[6]

This chapter exclusively briefs about the potential vulnerabilities of an IoT ecosystem and proposes a hybrid Key-based Twin VPN approach to address the Authentication and Authorization problems of a single node in any large scale IoT ecosystem in place. The chapter discusses briefly about the role of IoT in the routine life, components and architecture of a generic IoT model, hybridizing cloud technology with IoT, data security as a service of cloud based IoT, security challenges of IoT model, IoT – tampering the identity of the business or people, cybersecurity landscape of IoT, existing approaches, key generation for node authentication in IoT network, proposed Encrypted Twin VPN, advantages of Encrypted Twin VPN, performance analysis, conclusion, and future applications.

11.2   ROLES IN ROUTINE LIFE OF A HUMAN

It becomes possible to digitally have a control over the 99% of real-world objects and working environment that in turn evolves the next decade as the most advanced and automated digital era. Irrespective of the design and deployment issues, IoT offers lots of ease and comfort way of living in the globe by accordingly modernizing the things that support us in daily routines such as smartwatches that can monitor heartbeat and step counts, air-conditioners that can monitor the room-temperature and maintain it, microwave owens that can able to automatically cook for different types of food, to self-driving cars or planes that encompasses different types of complex sensors so as to identify optimal path and to detect various kinds of moving objects, a smart shoe that can sense real-time weight loss or gain and calories burnt which are vital to analyze and suggest suitable dietary and exercise plans custom-made to an individual.[1,2]

The only possible way to establish a seamless interconnection among the objects and devices is when such devices/objects can represent themselves digitally and henceforth they can be easily controlled remotely. The purpose of the connectivity then helps us fetch and inculcate more data from more places, ensuring an improved efficiency and security of the IoT devices.

The IoT platform acts as a transformational force that can help companies to improve their capabilities and enhance their performance through IoT analytics and IoT Security. In this decade, businesses in the utilities, oil and gas, insurance, manufacturing, transportation, infrastructure and retail sectors have already started to integrate AI, machine learning, and IoT together to make more informed decisions aided by the data aggregated and analyzed by the combo.[3]

11.3   COMPONENTS OF CLOUD ASSOCIATED IOT

The followings are the fundamental components of a generic IoT system:

1. **Sensors/devices:** Nodes that constitute sensor devices are considered as the basic unit of IoT which builds up the entire network whereas all the other devices are used to either aggregate or analyze the data perceived by these nodes as live feed from the surrounding environment. All these data are of different varieties and thus may have various levels of complexities to capture and analyze as well.

2. **Connectivity:** The aggregated data from such nodes were collectively sent to a centralized data repository where it can be efficiently stored, accessed, retrieved, and analyzed them on time. Since most of the IoT nodes are memory-constrained devices and being highly risky to cache data locally, they were either communicated to a local repository node capable of saving the data or communicate directly to the remote data server. In this case, it would be a cloud infrastructure and henceforth the IoT architecture has been designed like the data aggregated from every sensor ends up in the cloud repository using various mediums of communications such as LAN, PAN, MAN, WAN by deploying connection platforms like Zig-Bee, Wi-Fi, Bluetooth, etc.

3. **Data processing:** Once while the data stored either in the local repository or in a cloud, the customized software or machine learning AI applications perform the processing on the copy of the congregated data. This process can be a straight-forward and simple check like reading the temperature at that instant, ensuring the room temperature by using thermostats devices like AC or heaters. These types of networks are known as "underload" or "elementary" networks that deal with minimal data and can be processed within the local site itself. However, for complex tasks like detecting and identifying moving objects with the help of various computer vision image processing algorithms, an extensive and robust processing module is mandated. This is where ML algorithms and cloud computing plays a crucial role.

4. **User interface:** It is the most important component of the entire ecosystem since the intent of IoT is to smartly automate the environment that enhances the ease of living in such environment than before. The pivotal part of the component is that it communicates with the end-user. It can be a report or a notification seeking permission to automate or to get the feedback and utility of the event and so on. As IoT deals with a variety of data and the information or decision attained by the IoT system cannot be always readable or just triggers, the user sometimes might need an interface which can effectively interprets the underlying IOT system. For example, if an IoT module senses a rise in the room temperature than usual, instead of user, the module must be designed in such a way that it lowers the temperature of AC by itself else the purpose of IoT goes in vein.[4]

11.4   ARCHITECTURE OF IOT

The IoT is itself an ecosystem of a collection of smart devices and network devices that can able to transfer the data and communicate among them. Since the data are of higher volume and variety, it is only possible to analyze them to extract useful information and can attain a decision by coupling it with Big Data technologies and Cloud Computing (Fig. 11.2).

| Device Hardware | Device Software | Communications | Cloud Platform | Cloud Applications |

FIGURE 11.2   The IoT technology stack.

- **Sensing, embedded processing, connectivity:** The intent of the IoT ecosystem to actively sense its surrounding environmental factors such as temperature, gyroscope, pressure, etc. and is either process using the embedded coding in the devices or communicate to the higher layer of remote data storage over the networks such as Wi-Fi or Bluetooth or internet.

- **Smart devices and environment, cloud computing, and big data:** The perceived environmental data have been transferred or received through smart devices are then communicated to a centralized data server in a Cloud infrastructure which is then stored and processed as Big Data Analytics.
- **Technology, software, and application:** The IoT system uses varieties of connection protocols with respect to the underlying technologies and applications. The appropriate software for backend processing and frontend UI application has been designed in such a way that users have complete control over the communication and real-time connection among the various smart devices.
- **Users or groups of community:** The products or services generated by the IoT system are consumed by the users or communities so as to serve and lead a smarter life. The end-users can be an individual or an industry. It can even be the skeleton of a smart-city. As the beneficial group grows, the complexity of the system grows exponentially.[5]

11.5   HYBRIDIZING CLOUD WITH IOT

Cloud coupled IoT ecosystem can be considered as a kind of on-demand self-service. Since Cloud computing is a web-based service, it is possible now to control all the connected "things" at a cost of minimum sort of internet access to the IoT devices without waiting for any platform or special infrastructure.

Second, the cloud computing of IoT straightaway pushes the boundaries of the network access a broader and cheaper, meaning it offers several connectivity options. Without connecting it to an internetwork, the ultimate purpose of controlling the IoT devices fails down. Cloud computing makes it grant access to the resources at any-time any-where over a wide range of internet-enabled devices such as smartphones, pagers, smartwatches, tablets, desktops, and laptops. An end-user can access to the IoT ecosystem installed in the office or in parking lot or in home using a wide variety of devices having different architectures. This convenience is possible only by erecting robust network access points coupled with a secured internet.

Third, cloud computing had provisions for effective resource scheduling that makes aware of information can be shared among the legitimate users and nodes who have permission to access the particular resource by simply enforcing various SLA and Accession and Allocation policies. This approach makes it possible to establish wider association or tight connections with other nodes available in IoT ecosystem. It is possible by naming every "thing" (node) connected into the network can be uniquely identifiable by assigning a unique IP address and is now able to communicate with the other similar things and makes use of the resources and data distributed across the network in the ecosystem by making use of this IP.

11.5.1   DATA MODEL

The IoT technology has the potential to evolve and redefine the planet that enables everything can be seamlessly connected and are capable of communicating among each other so as to aggregate as many atomic and elementary environmental data to produce usable insight of the data and extract informational intelligent decisions. With the advent of adopting an IoT ecosystem in day-to-day routine of an individual, the physical world entities will acts as a huge pile of information systems that shares a common purpose of enlightening the quality and moral value of human life by empowering a variety of new arenas that make life smarter and simpler. As ML and AI makes things become self-dependent and autonomous, they are able to interact with other things through a communication gateway and make decisions to participate in the physical world by taking an appropriate and desired action. However, this seamless communication among the other nodes in an IoT also has a downslide while communicating more confidential and personal information across the common gateway that provides common access to any node in the same network that may offer an unwanted exploitable vulnerability. In case of a single vulnerable node in the entire security chain could possibly afford with limitless trapdoors that could potentially be exploited anytime and may lead data theft that includes the identity theft and information theft.

11.6   DATA SECURITY AS A SERVICE OF CLOUD COMPUTING BASED IOT

An open book truth is that cloud computing and IoT are tightly coupled in real-time. The emergence and the evolution of IoT and the rapid advancements of its associated technologies actually push the boundaries of making of "things" to be connected. As more nodes are already connected and billions to follow, it is going to be the mass production of tons of piles of data that is really very bigger in the history of Internet technology so far. Such enormous data have to be efficiently stored to promote rapid processing and accession. Since the backbone of any IoT system is its network, the need for database that works exceptionally well and fast over a network becomes the deciding factor. Thereby, Cloud computing is considered as an appropriate paradigm for storage and analytics of big data produced by IoT ecosystem. Despite being IoT itself already a ground-breaking technology that can transform the way of living, the actual stuff is that when it is hybridizing with the cloud computing.

The intent of hybridizing a cloud as a backend data storey and as a processing server of an IoT ecosystem enables a faster and secured data communication and a robust analyzing and processing of acquired data from the sensors that were connected to the cloud. Since IoT nodes are resource constraint nodes, it is very hard to deploy algorithms to achieve authentication, encryption or to accomplish decisions to disseminate apposite commands. The ultimate goal is to enable the transformation of such acquired information into a useful insight that in-turn produces cost-effective and more productive-based foreseen and corresponding action from those acquired insights. In the whole process, the ML and statistical algorithms that were already running in cloud space act as the brain of the IoT ecosystem that enables to yield an improved decision-making ability that can supports wider augmented internet-based applications.[7]

Cloud Computing emerges as a high-performance virtual infrastructure for deploying securing the system and data analyzing is made customizable and ease of use. One of the biggest advantages is that IoT becomes customer driven since the user interface with the actual IoT system can be accessed from anywhere and anytime while it is accessible through an internet which means there is no need to be present inside the range of the Bluetooth or Wi-Fi. While hybridizing IoT with the cloud computing repository and its services, a variety of new-fangled challenges ascend. It includes the delivery of quality services in time such as quality of service (QoS), end-user experience, data security, privacy, and reliability of the platform. Thus, the Cloud-based IoT affords a real-time utility-based solution to many businesses that needs 24×7 monitoring by provisioning users to access the business data through applications anytime and anywhere on-demand. And even, the user can take appropriate real-time measures in complete virtuality.

11.7   SECURITY CONCERNS OF IOT MODELS

Every connected device creates a vulnerable trapdoor for attackers in case of poor authentication schemes. The consequences of these vulnerabilities are extensive and devastative, even costs espionage and eavesdropping the entire network becomes possible. The risks posed include data spoofing, denial-of-device, denial-of-service, malfunctioning of devices, and so on. Many researches have been done in the past to address those risks and many security models were already in using to counter those risks. Despite, the main challenge of deploying IoT in terms of security remains associated with the security limitations of the low-cost and resource-constrained nodes that attract more attacks being vulnerable. The definition of a secured node in the IoT ecosystem may extents from the simplest policies to sophisticated designing measures. The Security policies and models should be on par with the course of vulnerabilities changes over time as threats evolve.

Beyond the resource constrained nodes of the system, other security issues ruins IoT:

- Unpredictable behavior—The huge number of deployed devices of variety of underlying architecture and independent designs that possess different set of enabling technologies making their behavior highly uncertain. A well-designed system that is in its bounded admissible control cannot assure about the behavior of such system while communicating with its other counterparts in the system.
- Device resemblance—Almost similar IoT nodes are fairly uniform and very difficult to identify if they have utilized the similar connection protocols and components. It is even more difficult to predict whether the entire system or a single device suffers from the vulnerability.
- Tricky deployment—Whenever it is deployed in tricky real-time situations or in a public-access domain, creates the biggest security issues of physically securing the devices as well the network since it is still the strange and earlier phase of security policies of the technology.
- No upgrade provision—Unlike many mobiles and Computers, smart devices are not designed to allow upgrades or any modifications since they were resource-constrained. And is certainly not possible after deployment into an IoT ecosystem.

- No security alerts—The major purpose of IoT that provides its implausible functionality without being blatant. There comes the most common problem like any other embedded technology of user awareness and consciousness about the nodes deployed. Since the users are not supposed to monitor the autonomous devices whether everything goes right. It is possible that a security breach can comfortably persevere over a long period of time.[8]

11.8 IOT—TAMPERING THE IDENTITY OF THE BUSINESS OR PEOPLE

IoT devices are intended to seamlessly monitor and grasp data from their environment that also includes people belongs to the environment. This is one of the greatest benefits and downside that introduces a perilous risk. Even though the abstract of the data may seem does not present any sort of risks involved, its subject and the potential issues after masquerading and sniffing over a prolonged exposure of the data stabs the deepest wound. Such activity can picturize a very clear identity and behavior of an individual, exposing to the criminals all such vital information can drags its danger over the advantages.[9]

This chapter discusses briefly about all the problems which are specific to IoT technology lead to many of its privacy issues, which primarily stem from the user's inability to establish and control privacy. Some of the vulnerabilities that promote such privacy issues are as follows:

11.8.1 THE RIGHT TO BE LEFT ALONE

The necessity of the security and the design of such security policies changes with time to time and is application dependent. It also depends on whether how the data are fetched, either the part of the data or the entire data itself and the user accession policies and the level of distribution of such data. The very important aspect is whether the IoT model handles the sensitive and confidential data like user login credentials like usernames, PINs, Passwords, or biometrics, etc. For example, if a user uses IoT model that opens the car door automatically by sensing the face, it is more dangerous to you to secure the sensors of the car or else either it opens for others or it will leak the feature sets of the face. It is like creeping into the privacy of an individual if the same sensor has been deployed under a public community like shopping malls because these sensors have the potential to monitor everyone's faces without their consent. IoT devices challenge these norms people recognize as the "right to be left alone."

11.8.2 INDISTINGUISHABLE DATA

IoT ecosystem can be deployed in a wide variety of ways with respect to the application and end-user convenience. In general, IoT implementation falls into group since the name itself speaks about "things" instead of "thing." And it is evident that only minimalistic data will be acquired for analytics from every node. To do so, the consent must be granted by the user to access all the nodes that were constituted the IoT ecosystem seamlessly. In this case, for each set of actions to be performed it needs to access every underlying devices of the ecosystem almost instantly and continuously, thus every IoT node must over tens of thousands requests simultaneously from a large pool of users for mere seconds in a public ecosystem.

11.8.3 GRANULARITY

The circumstances in which processing of big data (of different varieties and kinds) in a local repository leads a substantial vulnerability to the privacy of the data as well as nodes. IT gets even worse while this mode of data mining and analyzing combines with IoT ecosystem with its ability of granularity and considerably large in numbers. It attempts to sense the environment to extract the data almost continuously all the time or in a regular periodicals. There is a high probable risk of facilitating discrimination and exposing individuals to physical, financial, and reputation harm since the "things" already have a highly detailed profile.

11.8.4 CYBER ATTACKS

A possible vulnerability of a single IoT node is able to potentially expose the entire network to risk of various kinds of active and passive cyber attacks. Despite of the ability of the connections delivering prevailing combination and throughput, they also remain as the primary cause for the chaos like a compromised elevator or fire safety sprinkler system. The only measure to fix such defective chaos is to spot the exact defective node to fix the vulnerable points, which is arduous and the complexity of the task grows exponentially with the scale of the network.

11.8.5 DATA THEFT

The ability of acquiring the data by sensing a static or a dynamic environment continuously, aggregating and effectively storing in a local or cloud repository, analyses and extract meaningful insights and is transformed into actions or decisions and is finally transmitted to the concerned IoT node to initiate the action of IoT technology is its strength and its own pitfall. With this mighty ability to sense our environment, there is an evident possibility that if the security of such IoT system is compromised or breached, it is possible to easily access an individual's personal data to customized marketing/advertising with respect to their needs, identity theft, framing individuals for crimes, stalking, and much more bizarre.

11.9 EXISTING CYBER ATTACKS ON INTERNET OF THINGS

Privacy is not only the critical concern of IoT models, it is the same for all the technologies wherever information is shared over a network and the data are stored and analyzed later. Despite there were lots of efforts to make the firewall and security protocols more robust, the attackers and hackers always crafts a wonderful honey traps and backdoor applications to compromise them. As the technology evolves, the threats also getting more deadly and immutable to the security policies.

Many of the conventional routing protocols are too straight-forward, simple and thus more susceptible to vulnerabilities to recent evolved and some advanced real-time attacks of IoT networks. Some of them are as discussed below:

11.9.1 SPOOFING AND MANIPULATING ROUTING DATA

In this attack, the focus is to get a copy of all the data that were communicated over a particular router. Once then the illegitimate node can somehow managed to be a part of the IoT ecosystem and gained access to the centralized server, they were unstoppable since they were being able to regenerate the actual routing paths, create faulty messages, exploit the maximum end-to-end latency and much more.

11.9.2 SELECTIVE DENIAL-OF-SERVICE

In this threat, the malicious nodes may not forward either message that is generated by a particular node or drop all the messages that are destined to a particular node. It does not either interrupt the other nodes and usual data communication and it can also behave like a big black hole wherein it rejects all the message request to deny to process them or even acknowledge them; it simply swallows.

11.9.3 SINKHOLE ATTACKS

Unlike the above attack, by launching a sinkhole attack into the IoT network, the intent of the attack is to redirect all the traffic within a certain area to a particular compromised malicious node. It simply makes a particular node as the ultimate destination of all the data packets generated and destined to. The deadliest part of the attack is that it can also be able to create a faulty acknowledgement presuming from a legitimate node.

11.9.4 SYBIL ATTACKS

In this attack, a single node is cloned to possess the identities of all the other nodes that were connected in the IoT ecosystem. It is a very dangerous attack since it has the ability to nullify the effectiveness of any fault-tolerant systems by manipulating the packets of the geographic routing protocols itself. In this kind of attack, it is very hard to spot the malicious node since it seems to be everywhere in the network.

11.9.5 WORMHOLES

In the wormhole attack, an adversary in residing in a corner of the network who can receive messages over a low-latency link and keeps forwarding them to the other nodes of the network as a legitimate node. In this attack, there exist at least two different and independent malicious nodes that are detached from each other. One acts as a backup node for the other in one or many means.

11.9.6 HELLO FLOOD ATTACK

It is a modern attack that commonly affects all the wireless sensor nodes in a network where a node can be attached itself into the network easily by making a connection with any node by behaving lie its neighbor node of the network without any security breaches. This act of conjoining in the network is usually used to make the network high latency by sending fake packets and acknowledgements and keeps every other node busy all the time. It is made possible by simply flooding with the "Hello" packets to every other node.

11.9.7 ACKNOWLEDGMENT SPOOFING

In this attack, the attacker makes the other node to believe a dead node as highly active one or a weaker link to a highly optimal path to initiate the data transfer. In this attack, the attackers used to eavesdrop all the packets passing through the link. It becomes even deadly when it fakes the acknowledgment of the packets that have lost during the transmission.

11.10 KEY GENERATION FOR IOT NODES AUTHENTICATION

Some of the conventional measures against attacks are as follows:

- **Built-in security**—Both the individuals and the organization can adopt devices that have their own security policies already predefined in the drivers.
- **Encryption**—One of the successful and conventional approaches to keep the data secure is to encrypt and then transfer using PKI.
- **Risk analysis**—It is not only ends up in designing firewalls and security policies, but also review on regular periods to fix the modern vulnerabilities and update them to defend advanced threats and hacking approaches.
- **Authorization**—Authorization of nodes and users must be adopted in the model in almost wherever possible to enforce accession and

security policies as check point to ensure the nodes are legitimate and the data are authentic.

The basic idea of the encryption policy adopted in the chapter is to generate a 128 bit key for every session and is archived so that to make sure that it is not repeated and is useful in creating checksums.

ElGamal based Session key generation Algorithm has three phases such as key generation, encryption algorithm, and decryption algorithm (Fig. 11.3).

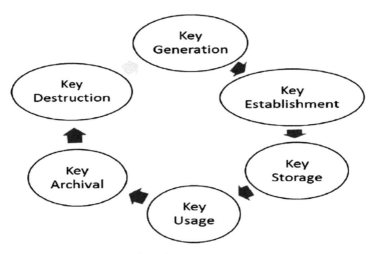

FIGURE 11.3   Key generation life cycle.

Key generation:

- Generate an efficient description of cyclic group G of order q with generator (this is a subset such that every element of group can be expressed in combination) g.
- Choose x randomly from 1 to q–1.
- Calculate h = gx.
- (G,q,g,h) is public key and x is kept as private key.

Encryption:

- Use receivers public key to encrypt message. Let it be (G,g,q,h).
- Choose y randomly from 1 to q–1.
- Calculate c1 = gy.
- Calculate s = hy.

- Map secret message m onto an element m' of G.
- Calculate c2 = m' × s
- Cipher text is (c1, c2) = (gy, m' × hy) where h = gx.

Decryption:

- Cipher text be (c1, c2) and private key be x.
- Calculate s = c1x
- Compute c2 × s−1, where c2 = m' × hy. This gives m'.c2 × s−1 = m' × hy * (g_{xy}) − 1 = m' × g_{xy} × g_{xy} = m'm' is the intended message.

Throughout the key lifecycle, secret keys must remain secret from all parties except those who are owner and are authorized to use them.

11.11 TWIN-ELGAMAL VPN

A dual VPN tunneling, popularly called as multi-hop VPN initially starts tunneling by encrypting the actual data and then pushes them into an unknown destined tunnel through a public or an insecure internet connection. The purpose of double VPN is to enhance the data security to the highest of level by thwarting unauthorized surveillance (Fig. 11.4). With this feature enabled, every bit of data that are being sending and receiving over the VPN connection is encrypted twice, with each encryption pass initiated at an independent point exclusively along its way. It is achieved by constantly manipulating the source and destination address at every point of encryption and is kept unanimous over a tunnel of infinite length from its either side.

To start off (and as it normally does), the device encrypts the data and sends it to the first VPN server in the chain. That server does what VPN servers usually do. It swaps out the actual IP address for its own and decrypts the data.

But, instead of forwarding the information right away to its ultimate destination, the first server encrypts it a second time. It then sends the re-encrypted data to a second VPN server.

That second server also does what VPN servers do. It undoes the encryption and replaces the source IP address (which is now that of the first VPN server) with its own.

Only at this point is the data forwarded to its final destination (say, a website you are viewing).

The key steps followed in the approach are as follows:

1. The raw data first get 128 bit encrypted and are designated the VPN server to reach the second VPN server in the chain which is the only known gateway to reach the destination and thus the destined VPN server acts as the destination node.

2. The second VPN server that has no chance trace beyond the source IP address which is the address of first VPN. IT ensures that it is absolutely no clue to even guess where the data have been originated from since the first VPN acts as a source.

On the whole, the dual VPN tunneling provides an enhanced security solution for both nodal authentication and conventional approach of encrypting at the nodal level that consumes memory and time. It acts as a one-stop solution for advanced network privacy and took security to the next dimension where the participants of the network protect from overloading and over-utilizing especially in energy, space, and time-constrained ecosystems and applications.

11.12   ENCRYPTED DUAL VPN

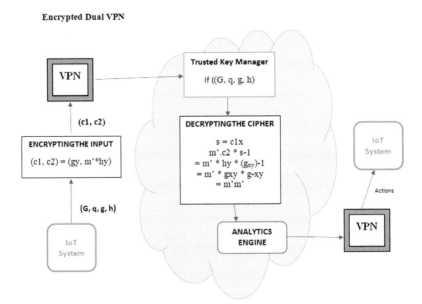

FIGURE 11.4   Architecture of twin encrypted VPN.

11.13 MANDATE OF DOUBLE VPN IN IOT ECOSYSTEM

As a security analysis, if the intent is to provide the security to the data transmitted over any network, a dual encryption is absolutely senseless. The fact is that the weaker deciphering attack such as a brute force attack would take about at least a billionth of billion years in order to compromise a data block that was encrypted by using a 128-bit key. In case of applying high-order attacks in heavy specification machine, it takes at least more than the time-pad to be elapsed for initiating a resending of the data. Thus, the encrypted data passing through a basic single VPN connection are already hardly possible to compromise.

The necessity comes in a real-time IoT environment especially that constitutes of a weakly resourced nodes and relatively unknown network where there is a possibility of threat not only to the data, but also to the entire IoT system. It is addressed only by enforcing strong authentication mechanisms and accessing policies which are not possible due to resource-constrained nodes and are impossible to authenticate every time the node sends data in a real-time monitoring system. For a time-constrained application such as Traffic management, Medical Nano-bots, real-time resource allocation, and management in pay-on-go cloud servers or in synchronizing an autonomous driving or pilot, it is not possible to spent time for authenticating nodes before the actual data transmission. The only option is to increase the data transmission mechanism considering both the aspects of hiding the identity to the other similar participants and to ensure highly secured network tunnel for every bit of data to be transferring.

Making the entire communication happen through two VPN servers drastically decreases the prospect of intercepting, snooping, dos, SQL injection, masquerading any legitimate node of the IoT ecosystem or the data communicating over the network.

11.14 ADVANTAGES OF A DOUBLE VPN

The utmost substantial contribution of a double VPN is its ability to ensure complete anonymity to both the nodes participating since both the VPN servers in the tunneling process have no chance of trace back each other using the actual IP addresses. The first server ensures that data are designated to another node through the virtual VPN address only. Both the communication happened through VPN's server as entry and exit points which leave even the ISP in a fuzzy. To ISP, the first node transmitted the data to the first VPN server and the rest is unknown beyond that point. It is to be understood that VPN always uses physical addresses of various geographic locations for every session that makes the attacker nearly impossible to track even the location of data originator.

A double VPN ensures the complete security to the network and the nodes that were participating in the data transmission in such a way that despite any of the VPN is used or compromised, it is almost not probable for anyone either to traceback to the originator or to intrude into the network.

11.15   PERFORMANCE ANALYSIS

The performance of the system has been evaluated in terms of average response time by simulating an IoT ecosystem having 100 active nodes that are static. Here, the average response time is calculated as the average latency time per number of real-time requests. The results were recorded by considering the cloud has been leased at the minimum of 2 GB dynamic memory with at least 20 MB cache with 2 Mbps dedicated bandwidth (Table 11.1).

TABLE 11.1   Average Response Time in MS.

S. No	Number of requests	Key-based authentication (KA) (in ms)	KA + single VPN (in ms)	KA + double VPN (in ms)
1	1000	140	152	175
2	2000	150	168	185
3	3000	165	179	192
4	4000	175	188	203
5	5000	200	215	235

It has been noted that though the average response time of the Dual VPN is greater than the Key based Authentication, it is acceptable with the respect to the complete anonymity and unbreakable encryption.

11.16   CONCLUSION

IoT provides a great platform for new business ventures and more importantly a better insight of ease living with the people that can considerably transform the common traits of humans and mode of living in the next decade. Hence, it is high-time to keep the infrastructure robust, secure, and trustworthy before it becomes commercial. In the chapter, the security to the data is seen as a part of securing the nodes and the entire infrastructure. It is believed that encryption in every single node is not possible in s resource constraint nodes and time- constrained applications. Since the security can be achieved 100% for the data communicated over a secured network, the scope of the extension of the work could be on the time that it consumes to provide the complete anonymity in the real-time network traffic.

Since the double VPN approach engaged dual encryption and dual tunneling of the data, it considerably requires the memory and time. Despite the time and memory consumption, for any hyper-conscious end-user who concerns about imperceptible network devices and a tamper-proof network either demanded by the application area or as a personal choice, a dual tunneling and encryption over VPN could be an optimal choice. With a trivial compromise in the response time, it is an absolute boon to not only keep the communication encrypted, but also the entire eco-system as anonymous and undiscoverable out of the network. As a future enhancement, efforts can be made to reduce the response time as well.

KEYWORDS

- **IoT**
- **VPN**
- **sensor**
- **security**
- **cloud computing**

REFERENCES

1. Lin, J.; Yu, W.; Zhang, N.; Yang, X.; Zhang, H.; Zhao, W. A Survey on Internet of Things: Architecture, Enabling Technologies, Security and Privacy, and Applications. *IEEE IoT J.* **2017,** 4 (5), 1125–1142.

2. Sharma, S.; Jeena, S. K. A Survey on Secure Hierarchical Routing Protocols in Wireless Sensor Networks. *Department of Computer Science and Engineering National Institute of Technology Rourkela,* Odisha, India, 2006.

3. Han, K.; Kim, K.; Shon, T. Untraceable Mobile Node Authentication in WSN. *Sensors* **2010,** *10* (5), 4410–4429.

4. Bellare and Rogaway. Entity Authentication and Key Distribution. *CRYPTO: Proceedings of Crypto* 1993.

5. Yang, Y.; Wu, L.; Yin, G.; Li, L.; Zhao, H. A Survey on Security and Privacy Issues in Internet-of-Things. *IEEE IoT J.* **2017,** 4 (5), 1250–1258.

6. Alharbi, S.; Rodriguez, P.; Maharaja, R.; Iyer, P.; Bose, N.; Ye, Z. FOCUS: A Fog Computing-based Security System for the Internet of Things. *IEEE Annual Consumer Communications & Networking Conference (CCNC)*, 2018, pp. 1-5, doi: 10.1109/CCNC.2018.8319238.

7. Stinson, D. R. *Cryptography: Theory and Practice*, 3rd ed.; Chapman and Hall/CRC: Boca Raton, 2005.

8. https://www.ibm.com/blogs/internet-of-things/what-is-the-iot/

9. https://tech.newstatesman.com/security/value-personal-data

CHAPTER 12

Role of Detection Techniques in Mobile Communication for Enhancing the Performance of Remote Health Monitoring

ARUN KUMAR[1*], MANOJ GUPTA[1], MOHIT KUMAR SHARMA[1], MANISHA GUPTA[2], LAXMI CHAND[1], and KANCHAN SENGAR[1]

[1]Department of ECE, JECRC University, Jaipur 303905, India

[2]Department of Physics, University of Rajasthan, Jaipur 302004, India

[*]Corresponding author. E-mail: arun.kumar1986@live.com

ABSTRACT

Today the human services biological system is looked with various difficulties running from framework, network, and asset. Gadget could have a reaction of few moments to hardly any milliseconds. The key would distinguish those viewpoints level of dormancy. For instance, progressed could require URLLC while wearable's intended for checking could do with request of seconds for answering to a focal cloud framework. While a great part of the exercises going on today, an exceptionally large walks will show up soon in the coming year. The cell administrators are engaging in the network access at the high information rate. The sensational improvement in the location plans of Orthogonal Frequency Division Multiplexing (OFDM), Multiple-Input Multiple-Output (MIMO), will assume a significant enhancement in remote medicinal services, as better network can accomplish in remote regions. In present day, remote health correspondence has procured wide scope of uses in various parts of human life. As much as the assortment of uses is expanding, the need of an ever increasing number of information rates is getting unavoidable with bigger volumes of information utilization.

The more seasoned ages of remote correspondence (like 2G and 3G) utilized system models dependent on sectored topology in which a radio wire component is engaged a specific way. The entire inclusion region is utilized to be isolated into various segments at specific points and every area is outfitted with a solitary receiving wire framework. In any case, this innovation has a constrained access to modest number of clients one after another. Consistently expanding request of higher information rates leaded the enthusiasm of research in various receiving wire frameworks, different transporter frameworks and later on, an innovation dependent on numerous bearer frameworks incorporated with remote health monitoring. Various MIMO Systems is a numerous reception apparatus framework in which both the transmitter and the beneficiary finishes are furnished with different receiving wire components. With the assistance of shrewd radio wire innovation and spatial decent variety MIMO Systems give information rates expanded numerous folds when contrasted with the traditional single receiving wire frameworks increasing performance of health monitoring in the remote areas. Single bearer frameworks require exceptionally huge transmission capacity for numerous receiving wire frameworks. In any case, remote health monitoring correspondence frameworks are constantly compelled of restricted data transfer capacity. To conquer this circumstance, various transporter frameworks are utilized in which a solitary bearer is partitioned into sub transporters and all the subcarriers are transmitted at once. OFDM is a multi-bearer innovation in which a solitary transporter is separated in various subcarriers which are symmetrical to one another. The Fourth Generation (4G) frameworks of remote correspondence use MIMO–OFDM System which gives enormous information transmission speed in a restricted data transfer capacity condition with better connection unwavering quality and improved inclusion zone, making it possible by sitting in one of the most advanced hospital and operating in the remote area. In this chapter, we have investigated just as depicted the numerous receiving wire frameworks (MIMO and Massive-MIMO) and their various perspectives with respect to framework model, structure and down to earth executions which will be useful to make better health monitoring in remote areas by reducing the detection time (delay). It is noted that, one can improve the connectivity in rural area by improving the detection schemes used in mobile communication. In the proposed chapter, Zero Forcing (ZF), Minimum Mean Square Error (MMSE) and Beamforming (BF) techniques are simulated and analyzed by using a Matlab-2014.

Execution of framework is investigated by figuring BER vs SNR. Based on results acquired, it is abridged that as the estimations of SNR expands BER diminishes for ZF and MMSE and it nearly disappears to zero in any event, for low SNR values if BF is utilized. In spite of the fact that ZF and MMSE are appropriate for planning a traditional MIMO System with four radio wire components. In light of the outcomes examined up until this point, it is concluded that BF can be utilized as a detection technique to enhance the radio transmission and reception performance in remote health areas.

12.1   INTRODUCTION

In the existing age, correspondence cells are not just utilized as a calling gadget, rather the utilization of cell gadgets is expanding exponentially consistently. A typical man utilizes advanced innovations for its day by day life needs like, transportation offices, medicinal offices, and correspondence offices. Innovation must be valuable to facilitate the life of individuals. Rapid information move systems are the essential prerequisite for innovation upgraded advanced world for development of administrations and henceforth the life of a person. To cook the need of huge measure of information at higher speed for ongoing applications cutting edge correspondence frameworks ought to be created. The remote frameworks can screen significant physiological parameters of the patients continuously, watch well-being conditions, surveying them, and generally significant, give input. Sensors are utilized in hardware medicinal and non-therapeutic gear and convert different types of imperative signs into electrical signs. Sensors can be utilized forever supporting inserts, preventive measures, long haul checking of impaired or sick patients. Social insurance associations like insurance agencies need ongoing, dependable, and precise demonstrative outcomes gave by sensor frameworks that can be observed remotely, regardless of whether the patient is in a medical clinic, facility, or at home.[1] The cutting edge remote systems known as 5G must have the option to address the limit limitations and the current difficulties related with current correspondence frameworks, for example, arrange interface unwavering quality, inclusion, dormancy, and vitality proficiency.[2] Personal satisfaction in many nations has been expanding much over the few hardly any decades because of huge enhancements in medication and opens medicinal services. Hence, there is an immense interest for the advancement of remote monitoring checking, which could be anything but difficult to use for older individuals. The remote medicinal services checking incorporates sensors, actuators, propelled correspondence advances and gives the open door for the patient to remain at his/her agreeable home rather in costly social insurance offices. In order to achieve better remote health monitoring, better detection techniques in Multiple-Input Multiple-Output (MIMO) based OFDM need to be realized. For high information rate and better connection unwavering quality exceptionally enormous number of radio wires (more than 64) at Base Transceiver Station (BTS) is utilized in monstrous MIMO. In enormous MIMO for countless reception apparatuses, this overhead will be exceptionally huge yet is overseen utilizing test-driven development (TDD) among uplink and downlink accepting channel correspondence which utilizes the early informations acquired from uplink pilots to be utilized in the downlink precoder. The down to earth execution of gigantic MIMO need synchronization with an enormous amount of autonomous radio frequency (RF) handsets and multiplication of information transports by a request for size or more which are extra difficulties to be experienced.[3] Monstrous MIMO framework establishes a cell coordinate with improved range and vitality proficiency.[4] Advantages of a massive-MIMO framework can be delighted in just when the exactness of CSI is kept up by both the base station and clients (that is at downlink and uplink both). A transmitter gets physical data of the patients, forms the information and sends through cell phones or video. At that point, the information is moved by the recipient to the PC. The specialist examines patient's data. On premise of data got, the specialist arranged in cutting edge clinic train the medicinal procedure. In the work, Zero Forcing (ZF), MMSE, and BF techniques are simulated and analyzed to increase the detection performance in the MIMO-OFDM for better remote health monitoring.

12.2 LITERATURE REVIEW

A ton of work has been carried out MIMO frameworks utilization in remote health monitoring. Be that as it may, useful acknowledgment of the framework requires various execution restricting difficulties to be tended to. For instance, in hypothetical system we think about uncorrelated diverts, however, in useful frameworks it is absurd to expect to keep up the partition required for rich dispersing spread condition among the reception apparatus components in a huge cluster and an outcome we get corresponded channels. Another test in pragmatic acknowledgment is to build up a monetarily practical system with ease and low-power RF segments. Essentially, remote channels show recurrence particular blurring, which requires expensive RF segments. Monstrous MIMO joined with OFDM is the response to this recurrence particular issue and will decrease the monetary weight of significant expense RF segments. Monstrous MIMO innovation is a future trust in cutting edge remote correspondence frameworks past 4G (fourth era correspondence frameworks). At present, 4G utilizes multi-client MIMO frameworks which have roughly equivalent number of administration receiving wires and dynamic terminals. Hoydis et al. suggested that the exhibition of broadcast of a gigantic framework, with a general channel model dependent on arbitrary network hypothesis, is equal for coordinate channel (MF), a straightforward locator and a base min-square-blunder (MMSE) identifier gave the quantity of base station radio wires is exceptionally huge accessible by air is huge adequate with a viable SNR. This outcome is helpful for investigation of downlink and progressively reasonable channel models too.[5] Lee et al. proposed another sort of monstrous MIMO frameworks named as network-gigantic MIMO frameworks in which three broadcasting units are associated with a various client terminals at the cell-limit in a similar recurrence band.[6] Constrained participation among RUs chooses the extent of this framework in down to earth purposes. Bjornson et al. proposed a system, empowered with higher bandwidth reuse, which is useful in improving the vitality proficiency of a huge MIMO framework to limit the all-out force utilization (static equipment force and dynamic radiated power both) without bargaining for performance. The enormous MIMO framework is functional in little cells utilizing ideal and low intricacy non-intelligent pillar shaping from numerous transmitter receiving wires prepared at base stations.[7] Mohammed and Larsson recommended that in a recurrence specific multi-client gigantic MIMO framework, to keep up the independent of the channel gain and the data images to be imparted, a precoding calculation must be utilized to create constant-envelope (CE) signals at each base station with exceptionally low intricacy.[8] Bjronson et al. expressed that a vitality proficient gigantic MIMO framework can be intended for high SNR utilizing obstruction stifling precoding plans, similar to ZF precoding, for countless radio wires sent at base station to serve numerous clients all the while.[9] Zhu et al. recommended that in a multicell enormous MIMO framework a verified downlink transmission can be structured utilizing match channel precoding and counterfeit noise (AN) age at the base station dependent on arbitrary. For the streamlined force designation for AN age, the rate is just an expanding capacity of the quantity of base station radio wires in a polluted pilot condition.[10] Zhang et al. dissected the uplink execution of a monstrous MIMO framework, for MRC and ZF recipients and for Ricean blurring channels and inferred that for a similar beneficiary, the data for great and flawed CSI will be a set an incentive for an expanding Ricean K-factor.[11] Truong and Heath[12] recommended that the impact of channel maturing on uplink and downlink attainable rates for a match filtering precoder shows that channel maturing mostly influences the ideal sign capacity to a client and furthermore bury cell obstruction because of pilot sullying. To the extent downlink channel estimation is a thought of it as, requires huge overhead caused because of countless radio wires at base station which brings about constrained execution (restricted information paces) of massive-MIMO frameworks.[12] Hoydis et al.[13] suggested that in a TDD-based monstrous MIMO framework CSI can be acquired utilizing channel correspondence by enabling each gadget to reuse it got obstruction covariance grid gauge for precoding in an impedance mindful condition which is reliant on nearby data just, with no information trade between different gadgets in the system.[13] Choi et al. proposed two diverse preparing plans for downlink FDD gigantic MIMO framework, in particular, open-circle and shut circle preparing structures dependent on progressive forecast of the channel at the client end. In contrast with open-circle system, channel estimation execution can be improved with a little length of preparing signals in each blurring squares in a shut circle preparing structure by utilizing the past got preparing signals at client end as input. Therefore, a diminished overhead for downlink preparing is gotten which devours less transmission capacity which is a valuable asset and an imperative for cell correspondence frameworks. Regular CSI estimation approaches are not reasonable for FDD huge MIMO frameworks because of the enormous preparing and input overhead for downlink. Choi proposed a calculation to decrease input overhead for CSI of FDD-based monstrous MIMO frameworks, which is named as receiving wire bunch beam forming (AGB). A gathering of connected reception apparatuses having comparable examples are mapped to a solitary representative worth. This gathering based mapping decreases the element of channel vector. A code-word is picked for each representative incentive from the codebook produced by new decreased measurement channel vector. In this manner, the input data for CSI are diminished considerably by conveying it for new emblematic incentive as opposed to imparting for singular reception apparatus components.[14] Masood et al. recommended the calculation dependent on this methodology can discover estimation in any event, for channels by allotment of the data of neighboring receiving wires for dynamic system by synchronization between the reception apparatuses in the system, therefore, the need of countless pilots is diminished.[15] Bazzi et al. proposed a two-arrange precoding plan for the MIMO radio wires and of transmitter/beneficiary matches regardless of the extent of broadcasting reception apparatuses.[16] Zhao et al. recommended that channel measurements of MIMO frameworks got from dynamic channel models and channel spatial connection models are utilized to structure ideal preparing arrangement dependent on successive data estimation.[17] Gao et al. proposed a FDD massive-MIMO framework with the assistance of block orthogonal for utilizing non-symmetrical pilots (ordinarily symmetrical pilots are utilized).[18] Channel estimation is considerably debased by pilot defilement and accordingly the exhibition of gigantic MIMO frameworks is constrained seriously by corruption in channel estimation caused because of pilot sullying. To acquire the channel estimation execution like an obstruction free condition, Yin et al. proposed a covariance-supported channel estimation strategy dependent on Bayesian estimator which totally evacuates the impact of pilot sullying.[19] Bogale and Le proposed a pilot enhancement for a TDD based multiuser gigantic MIMO framework. Proposed calculation initially details channel estimation issue as weighted total MMSE containing pilot images and some presented factors. Later, these presented factors are enhanced after utilizing MMSE equalizer. Pilot images are enhanced utilizing semi clear programming (SDP) curved advancement for equivalent number of UE reception apparatuses and pilot images. Re-enactment results show that proposed calculation decreases the pilot defilement when contrasted with the channel estimation calculation dependent on MMSE. Client limit in downlink correspondence in a TDD massive-MIMO framework is restricted by pilot pollution as it makes the signal-to-interference-plus-noise ratio (SINR) soaked.[20] Bjornson et al. suggested that equipment disabilities at handsets make non-zero estimation mistake floors regardless of the power and the quantity of base receiving wires, which is absolutely as opposed to the results got by the utilization of perfect equipment. Stage commotion is a significant disadvantage of equipment impedances made because of the utilization of low linearity power speakers.[21] Pitarokoilis et al. considered the impact of stage commotion on recurrence specific gigantic MIMO frameworks with defective CSI for two unmistakable activity modes in particular, synchronous, and non-synchronous tasks. Scientific outcomes appear as opposed to utilizing completely synchronous commotion sources utilization of autonomous sources is gainful. At base station, for recognition the data images combiner is utilized.[22] Xu et al. considered the enormous MIMO Full Duplex Relaying (FDR), outfitted with exceptionally huge receiving wire exhibits, to decide the impact of equipment disabilities in regards to obstructions and successful bends and reasoned that the presentation of monstrous MIMO framework is restricted by the equipment impedances at sources and goals (i.e., at client types of gear) and not by the transfer, which unmistakably demonstrates that base stations can be furnished with non-perfect minimal effort equipment with no tradeoff with the exhibition of the framework. To make the massive-MIMO framework financially suitable utilizing ease and low force RF frameworks Bannour.[23] Shen et al. made a broad investigation about the common sense transmission plans which proposed that the massive-MIMO-OFDM downlink methods are a lot of reasonable for execution in handy situations as it abuses the brace of receiving wires and utilizes the carriers rather than a solitary bearer for information broadcasting.[24] Shen et al. did similar examination dependent on interface level recreation of enormous MIMO-OFDM in nearness of non-straight twisting of intensity enhancers with exceptionally huge number of transmitter reception apparatuses and presumed that throughput of monstrous MIMO-OFDM is higher than that of huge MIMO SC-FDMA. Channel estimation is a major test for gigantic MIMO-OFDM framework in versatile multipath channels.[25] To determine this issue, Love et al. proposed a TFT–OFDM transmission conspires for enormous OFDM framework where time-recurrence preparing data for each TFT-OFDM image without CP is established when space preparing arrangement time-space (TS) together with the recurrence area symmetrical gathered pilots. At the beneficiary, way postpone estimation, without the impedance wiping out, is finished utilizing the TS, while recurrence area pilots are utilized to figure out the way gains and therefore channel variety is effectively determined utilizing this time-recurrence joint channel estimation conspire. Recreation results show that when contrasted with the basic monstrous MIMO-OFDM frameworks, the proposed TFT-OFDM gigantic MIMO plot has significantly more range productive and furthermore having coded bit mistake rate execution practically equal to the ergodic divert limit in versatile environment. The strategy utilized for CSI estimation intensely influences the presentation of enormous MIMO-OFDM framework. Enormous overhead delivered because of the colossal reception apparatus cluster at base station lessens the Multiple Access Channel (MAC) proficiency.[26] By utilizing an uncommon pilot structure and unitary framework criticism, an exceptional strategy for early detection was proposed by Kudo et al. which empowers the passage to get CSI estimation for all transmitting radio wires utilizing a couple of preparing outlines. The utilization of OFDM makes high peak normal force proportion (PAPR) which makes enormous MIMO-OFDM framework costly as it needs refined RF segments.[27] Prabhu et al. proposed a PAPR decrease strategy, named as Antenna-Reservation technique, having a low multifaceted nature overhead and which can be actualized with basic equipment like a DFT square and a precoder. In this technique, signals sent to lots of receiving wires are cut essentially and to repay this remedy, signals are transmitted to lots of saved radio wires (reception apparatus reservation). Re-enactment results show that 25% saved receiving wires can decrease the PAPR by 4 dB.[28] Thus, higher linearity prerequisites of RF parts at base station are decreased and the MU-massive-MIMO-OFDM framework can be comprised with the minimal effort RF types of gear at the base station. It is accepted, hypothetically, that MIMO directs are free yet practically speaking, for huge number of receiving wires, channels are connected on the grounds that it is hard to isolate enormous number of reception apparatuses appropriately. Fang et al. contemplated massive-MIMO under corresponded channels and watched the presentation of the framework dependent on their recreation results. The presentation of the framework is more awful with increment in connection coefficient of channels. In the event where the estimation of channel connection coefficient is littler than a specific little worth, the presentation will upgrade rapidly with increment in number of base station reception apparatuses and for channel relationship coefficient with bigger incentive than the specific little worth the exhibition of massive-MIMO is improved gradually with increment in number of receiving wires at base station.[29] To improve the presentation, Li et al. proposed three-dimensional MIMO (3D MIMO) with dynamic Beamforming (BF), for monstrous MIMO frameworks, in which receiving wire components are conveyed in the two measurements that is evenly just as vertically at transmitter and recipient the two finishes. In proposed dynamic BF calculation, UEs' level and vertical the two bearings are considered to ascertain the BF vector as opposed to taking just even heading (as if there should be an occurrence of customary MIMO). In 2D BF strategy, UEs at cell limit are vigorously influenced by entomb cell obstructions. The proposed dynamic BF calculation decreases the intercell obstructions in a viable way and therefore a sharp increment in throughput of the UEs in the whole cell, including the cell limit and its inside is watched.[31] Bogale proposed half-breed BF for downlink multiuser massive-MIMO frameworks. In the proposed BF method, it is expected that ideal CSI is accessible at transmitter and it is viewed as an all-out entirety rate amplification issue. The half-breed BF (blend of simple and computerized BF) depends on gauged entirety mean square blunder (WSMSE) minimization issue which is understood by applying the hypothesis of compacted detecting. The exhibition of crossover BF is practically equivalent to the computerized BF if the quantity of multiplexed images is less. Yue and Li proposed transmit and get conjugate BF which is a viewable pathway (LOS) based transmission plan to dispose of issues created by polluted pilots. It decreases the overhead required for full CSI. In the proposed conjugate BF precoding plan, dissipated part of the sign is considered as impedance over Rician level blurring channel condition. It is obvious from re-enactment results that the expansion in transmit power prompts disappear the dissipated obstructions and accordingly better ergodic rate is accomplished when contrasted with the MMSE-based channel estimation. Numerical outcomes for the proposed conjugate BF plot show that the effect of connected radio wire components is not that much seriously affecting. In this way, enormous number of reception apparatuses can be set minimalistically at the base station. ZF-Beam shaping (ZF-BF) is a reasonable innovation for complete between pillar obstruction wiping out in a multiuser-MIMO framework; however, it is not appropriate to execute in the massive-MIMO framework because of its tremendous handling unpredictability for enormous number of base station receiving wires. Park et al. proposed a ZF-BF conspire with decreased unpredictability dependent on consecutive between pillar impedance retractions in a plummeting request of their quality. Numerical and recreation results show that as the quantity of between pillar impedance undoing builds, the presentation of the proposed plan shifts toward that of ZF-BF. Finally, it is up to 90% of limit of ZF-BF with 2–7% multifaceted nature of ZF-BF for countless receiving wires (state 128). Lakshminarayana et al. proposed a multicell BF calculation for enormous MIMO frameworks, in view of irregular lattice hypothesis, with a constrained data trade about the channel insights rather than complete CSI between the BSs, to limit the necessity of absolute transmit power for each base station. Rather than limiting total transmit power for all the base stations, this proposed calculation is fit for limiting the transmit power for singular base stations. Reproduction results show that proposed calculation is fit for accomplishing the required SINR and furthermore considerable force sparing when contrasted with zero constrained BF (ZFBF) when a number of radio wires per base station are equivalent arranged by greatness to the quantity of UEs in every phone for the monstrous MIMO frameworks. In a massive-MIMO framework, for precoding and location, ZF strategy is considered as for all intents and purposes near the channel limit of downlink and uplink. For countless base station, radio wires (state M) and nearly less number of clients (state K, where M >> K) equipment usage of ZF are particularly perplexing because of the intricacy of figuring the backward of the K × K lattice.

12.3 SISO VS MIMO

An extremely basic inquiry emerges here: why MIMO is viewed as superior to ordinary SISO framework? We have attempted to answer it with the assistance of re-enactment model. Single information single yield (single input, single output, SISO) framework is regular cell correspondence framework, in view of sectorized topology, is outfitted with single radio wire at collector end and a solitary reception apparatus at the transmitting end. It has an extremely restricted limit with respect to information rate, connect dependability and number of client terminals, and so on then again, MIMO framework is furnished with various receiving wire at recipient and transmitter both the finishes. Its channel limit is far superior to the customary SISO framework. A reproduction result dependent on the scientific model of SISO and MIMO (16 × 16) is shown in Figure 12.1 which obviously delineates the limit upgrade given by the enormous scale MIMO frameworks.

12.4 SIGNAL DETECTION

In a different reception apparatus framework at the recipient, it is important to utilize some appropriate system to locate the ideal sign for singular radio wire component out of the envelope of got signals. This procedure is known as sign identification. For this sign, identifiers must be utilized at the recipient. An assortment of sign recognition strategies is accessible for instance, straight sign locators, and edge-based sign indicators.

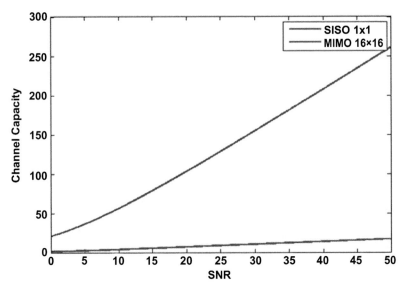

FIGURE 12.1   Capacity vs SNR.

12.4.1   *LINEAR SIGNAL DETECTORS*

In a straight sign location plot, all the transmitted signs, aside from the ideal information stream from the objective transmit reception apparatus, are treated as impedances. To distinguish the ideal sign from the point to broadcasting the wire at accepting end, all the impedance data are invalidated or limited utilizing some reasonable procedures. ZF and MMSE identification systems are two standard straight sign discovery methods.

12.4.2   *ANGLE-BASED SIGNAL DETECTORS*

Shaft shaping (BF) is a standard point based sign discovery plot. In shaft framing signal location procedure, we utilize close-circle transmit decent variety conspire where channel is now known to the recipient.

12.4.3 ZERO FORCING SIGNAL DETECTION

ZF is a direct sign identification plot. All the transmitted signs, with the exception of the ideal information stream from the objective transmit receiving wire, are treated as obstructions in a straight sign discovery plot is indicated in Figure 12.2. To identify the ideal sign from the transmit receiving wire at accepting end, all the obstruction signs are invalidated utilizing some reasonable systems.

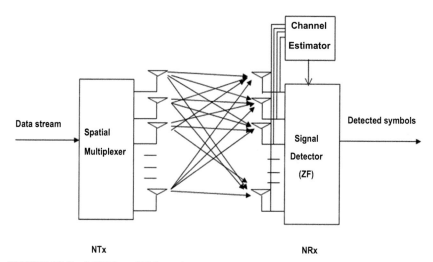

FIGURE 12.2   MIMO on ZF detection.

12.4.4 MMSE SIGNAL DETECTION

To recognize the ideal sign from the radio wire at accepting end, all the impedance signals are limited utilizing some reasonable systems is indicated in Figure 12.3.

12.4.5 BEAMFORMING

In this scheme, recognition strategy utilizes close-circle remit assorted variety plot, where channel is now known to the recipient indicated by Figure 12.4. In this technique, beneficiary remits criticism about the channel data to the objective spreader.

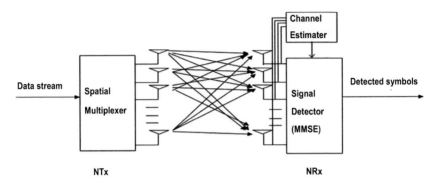

FIGURE 12.3   MMSE detection.

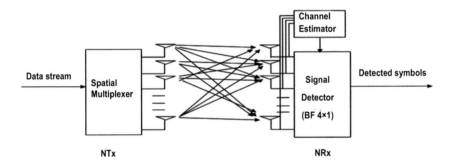

FIGURE 12.4   Beamforming.

12.5   SIMULATION RESULTS

In view of the hypothetical and scientific models proposed in past segments, a regular MIMO framework is mimicked utilizing MATLAB. Recreation results for figuring of BER of a 4 × 4 MIMO framework utilizing diverse sign location procedures are recorded.

12.5.1   *BER OF ZF DETECTION*

The bit error rate (BER) of 4 × 4 schemes using ZF technique is given in Figure 12.5 and the tabular value is represented in Table 12.1.

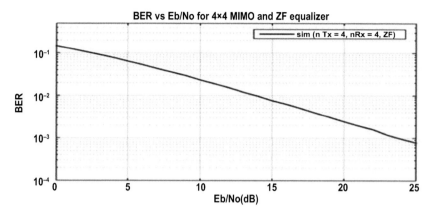

FIGURE 12.5 BER vs Eb/no.

12.5.2 BER OF MMSE

The BER for 4×4 using MMSE equalization is given in Figure 12.6 and the tabular result is indicated in Table 12.2.

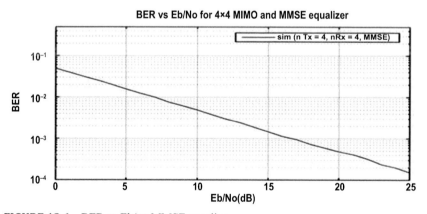

FIGURE 12.6 BER vs Eb/no MMSE equalizer.

12.5.3 BEAMFORMING OF BER OF A 4×1 MIMO

The 4×1 MIMO framework, recreation results for BER utilizing BF leveling are shown in Figure 12.7 and the tabular representation is indicated by Table 12.3.

TABLE 12.1 BER vs Eb/No.

Eb/No(dB)	0	1	2	3	4	5	6	7	8	9	10	11	12
BER	0.146756	0.127177	0.10844	0.092102	0.077729	0.064321	0.05329	0.043349	0.035631	0.029437	0.023265	0.01901	0.015285

Eb/No(dB)	13	14	15	16	17	18	19	20	21	22	23	24	25
BER	0.011955	0.009732	0.007589	0.00623	0.004962	0.003906	0.003107	0.002434	0.001947	0.001591	0.001178	0.000952	0.000776

TABLE 12.2 BER vs Eb/No.

Eb/No(dB)	0	1	2	3	4	5	6	7	8	9	10	11	12
BER		0.04902	0.03951	0.03136	0.02536	0.019964	0.015594	0.012387	0.010034	0.007569	0.006132	0.004913	0.003812	0.00296

Eb/No(dB)	13	14	15	16	17	18	19	20	21	22	23	24	25
BER		0.00242	0.001873	0.001462	0.001127	0.000947	0.000729	0.000594	0.000481	0.000411	0.000323	0.000235	0.000199	0.000152

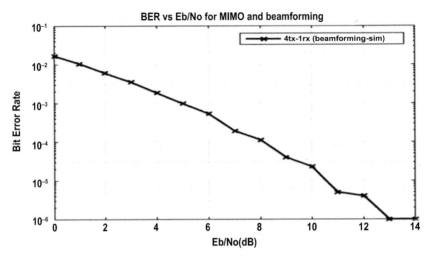

FIGURE 12.7   BF equalizer for 4 × 1 MIMO using.

12.5.4   BER SIMULATION RESULTS FOR 4 × 4 MIMO

Re-enactment outcomes for 4 × 4 MIMO, utilizing three distinctive adjustment procedures (ZF, MMSE, and BF), are examined to get their exhibitions dependent on BER. Zero constraining adjustment attempts to invalidate the obstruction. It gives a modest BER esteem for superior SNR. It is anything but difficult to actualize for all intents and purposes due to its less multifaceted nature. The reproduction results demonstrate that to acquire a superior presentation MMSE is similarly better procedure for the post-recognition balance. Be that as it may, MMSE is nearly mind boggling to structure and execute for all intents and purposes. One more method called BF is utilized here to limit the impedances with the goal that the framework execution is expanded. Reenactment results show gigantic outcomes for BF. For higher SNR, very nearly zero BER is watched for huge number of transmitted bits. It might alter less in viable condition, however, it gives the most ideal BER. For an enormous number of reception apparatuses at base station, BF is increasingly helpful to structure and execute utilizing a DSP processor. We will attempt to improve the present 4 × 4 MIMO model to additional levels with an ever increasing number of radio wires at base station utilizing BF strategy in our future work. In light of the reproduction results acquired in regular MIMO frameworks, we

TABLE 12.3 BF Equalizer for 4 × 1 MIMO Using.

Eb/No(dB)	0	1	2	3	4	5	6	7	8	9	10	11
BER	0.018816	0.01044	0.005995	0.003543	0.001857	0.000988	0.000537	0.000193	0.000113	4.00E-05	2.30E-05	5.00E-06

Eb/No(dB)	12	13	14	15	16	17	18	19	20	21	22	23	24	25
BER	4.00E-06	1.00E-06	1.00E-06	0	0	0	0	0	0	0	0	0	0	0

proposed 8 × 1 MIMO framework utilizing BF signal recognition plan to get the improved throughput when contrasted with the present framework (4 × 1 MIMO). For the 8 × 1 MIMO framework, recreation results for BER utilizing BF adjustment are given in Figure 12.8 and tabular value is given in Table 12.4.

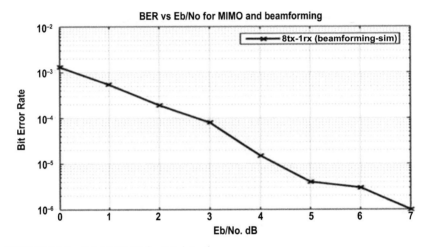

FIGURE 12.8   8 × 1 MIMO for BF detection.

For an 8 × 1 MIMO system, BER vs SNR with and without BF detection schemes is shown in Figure 12.9 and tabular value is given in Table 12.5. In this chapter, we examine the various detection techniques in healthcare system and its adoption in 5G technology. Re-enactment results show that BER is significantly decreased even at smaller SNR in the event that we use BF. On the off-chance that various receiving wire frameworks are utilized without the BF location method. The recreation results show an alternate sort of result. BER is a lot or more prominent even at moderate and higher SNR. A relative investigation of forbidden qualities shows that the framework execution is improved significantly with the utilization of BF method.

12.6   CONCLUSION

The propelled recognition procedures will bring a more significant level of cooperative energy and progressively innovative advances. These incorporated

TABLE 12.4 8 × 1 MIMO for BF Detection.

Eb/No (dB)	0	1	2	3	4	5	6	7	8	9	10	11
BER	0.001314	0.000536	0.000191	8.00E-05	1.50E-05	4.00E-06	3.00E-06	1.00E-06	0	0	0	0

Eb/No (dB)	12	13	14	15	16	17	18	19	20	21	22	23	24	25
BER	0	0	0	0	0	0	0	0	0	0	0	0	0	0

systems will stretch out data giving a preferred position to control nature. Recreation outcomes show that BER is extensively diminished at lesser SNR on the off-chance that we use BF. In the event that numerous reception apparatus framework is utilized without the BF identification system; the reproduction results show an alternate sort of result. BER is a lot more prominent even at moderate and higher SNR. A similar investigation of forbidden qualities shows that the framework execution is improved significantly with the utilization of the BF system. Hypothetically, we can get an awesome framework execution; however, there are a great deal of issues in viable usage like, unpredictability of equipment structure because of huge number of radio wires (which are associated), equipment weaknesses, and limit of computerized signal processor, continuous preparing of the signs, power requirements, and cost of the supplies. For additional enhancements, an exhaustive report is required to defeat these usage related issues.

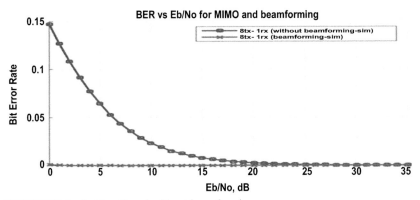

FIGURE 12.9   8 × 1 with and without beamforming.

TABLE 12.5 8 × 1 with and without Beamforming.

Eb/No (dB)	0	1	2	3	4	5	6	7	8	9	10	11
BER	0.147276	0.126864	0.10815	0.091809	0.077087	0.064507	0.052804	0.043422	0.035469	0.028491	0.023123	0.01889

Eb/No (dB)	12	13	14	15	16	17	18	19	20	21	22	23	24	25
BER	0.01486	0.01225	0.0097	0.00775	0.00620	0.00499	0.00392	0.00325	0.00241	0.00199	0.00164	0.00125	0.00096	0.00078

Eb/No	26	27	28	29	30	31	32	33	34	35
BER	0.000652	0.000536	0.000425	0.000309	0.000232	0.000192	0.000164	0.000104	9.60E-05	6.30E-05

KEYWORDS

- **health monitoring**
- **detection schemes**
- **4G**
- **MIMO**
- **BER**

REFERENCES

1. Kumar, K. Design and Simulation of MIMO and Massive MIMO for 5G Mobile Communication System. *Int. J. Wireless Mob. Comput.* **2018,** *14,* 197.
2. Marzetta, T. L. Noncooperative Cellular Wireless with Unlimited Numbers of Base Station Antennas. *IEEE Trans. Wireless Commun.* **2010,** *9* (11), 3590–3600.
3. (http://www.ni.com/white-paper/52382/en/).
4. Liu, W.; Han, S.; Yang, C. Energy Efficiency Comparison of Massive MIMO and Small Cell Network. In *IEEE Global Conference on Signal and Information Processing (GlobalSIP)*; 2014; pp 617–621.
5. Hoydis, J.; Ten Brink, S.; Debbah, M. Massive MIMO: How Many Antennas Do We Need? In *IEEE 49th Annual Allerton Conference on Communication, Control, and Computing (Allerton)*; September 2011; pp 545–550).
6. Lee, C.; Chae, C. B.; Kim, T.; Choi, S.; Lee, J. Network Massive MIMO for Cell-boundary Users: From a Precoding Normalization Perspective. In *Globecom Workshops (GC Wkshps)*; IEEE, December 2012; pp 233–237.
7. Bjornson, E.; Kountouris, M.; Debbah, M. Massive MIMO and Small Cells: Improving Energy Efficiency by Optimal Soft-cell Coordination. In *20th International Conference on Telecommunications (ICT)*; IEEE, May 2013; pp 1–5.
8. Mohammed, S. K.; Larsson, E. G. Constant-envelope Multi-user Precoding for Frequency-selective Massive MIMO Systems. *Wireless Communications Letters, IEEE* **2013,** *2* (5), 547–550.
9. Bjornson, E.; Sanguinetti, L.; Hoydis, J.; Debbah, M. Designing Multi-user MIMO for Energy Efficiency: When Is Massive MIMO the Answer?. In *Wireless Communications and Networking Conference (WCNC)*; IEEE, April 2014; pp 242–247.
10. Zhu, D.; Li, B.; Liang, P. Matrix Inversion Approximation Based on Neumann Series in Massive-MIMO Systems. *IEEE ICC—Wireless CommunicationSymposium,* 2015, 1763–1769.
11. Zhang, Q.; Jin, S.; Wong, K. K.; Zhu, H.; Matthaiou, M. Power Scaling of Uplink Massive MIMO Systems with Arbitrary-Rank Channel Means. *Selected Topics Sign. Process. IEEE J.* **2014,** *8* (5), 966–981.

12. Truong, K. T. and Heath, R. W. Jr. Effects of Channel Aging in Massive- MIMO Systems. *J. Commun. Netw.* **2013**, *15* (4), 338–351.
13. Hoydis, J.; Hosseini, K.; Brink, S. T.; Debbah, M. Making Smart Use of Excess Antennas: Massive MIMO, Small Cells, and TDD. *Bell Labs Tech. J.* **2013**, *18* (2), 5–21.
14. Choi, J.; Chance, Z.; Love, D. J.; Madhow, U. Noncoherent Trellis Coded Quantization: A Practical Limited Feedback Technique for Massive MIMO Systems. *IEEE Trans. Commun.* **2013**, *61* (12), 5016–5029.
15. Masood, M.; Afify, L. H.; Al-Naffouri, T. Y. Efficient Coordinated Recovery of Sparse Channels in Massive MIMO. *IEEE Trans. Sign. Process.* **2015**, *63* (1), 104–118.
16. Bazzi, S.; Dietl, G.; Utschick, W. Subspace Precoding with Limited Feedback for the Massive MIMO Interference Channel. In *Sensor Array and Multichannel Signal Processing Workshop (SAM)*, 8th; IEEE, June 2014; pp 277–280.
17. Zhao, Y.; Wang, X.; Gu, X.; Wan, W.; Pang, Q. Training Sequence Design for Channel State Information Acquisition in Massive MIMO Systems. *Personal, Indoor, and Mobile Radio Communications (PIMRC), IEEE 26th Annual International Symposium*, 2015; pp 1712–1716.
18. Gao, Z.; Dai, L.; Dai, W.; Wang, Z. Block Compressive Channel Estimation and Feedback for FDD Massive MIMO. *Computer Communications Workshops (INFOCOM WKSHPS), 2015 IEEE Conference*, 2015; pp 49–50.
19. Yin, H.; Gesbert, D.; Filippou, M. C.; Liu, Y. Decontaminating Pilots in Massive MIMO Systems. In *2013 IEEE International Conference on Communications (ICC)*, June 2013; pp 3170–3175.
20. Bogale, T. E.; Le, L. B. Pilot Optimization and Channel Estimation for Multiuser Massive MIMO Systems. In *Information Sciences and Systems (CISS), 2014 48th Annual Conference on*, March 2014; pp 1–6.
21. Bjornson, E.; Hoydis, J.; Kountouris, M.; Debbah, M. Hardware Impairments in Large-scale MISO Systems: Energy Efficiency, Estimation, and Capacity Limits. In *18th International Conference on Digital Signal Processing (DSP), IEEE*, July 2013.
22. Pitarokoilis, A.; Mohammed, S. K.; Larsson, E. G. Uplink Performance of Time-reversal MRC in Massive MIMO Systems Subject to Phase Noise. *IEEE Trans. Wireless Commun.* **2015**, *14* (2), 711–723.
23. Xu, K.; Gao, Y.; Xie, W.; Xia, X.; Xu, Y. Achievable Rate of Full-Duplex Massive MIMO Relaying with Hardware Impairments. *Communications, Computers and Signal Processing (PACRIM), IEEE Pacific Rim Conference*, 2015; pp 84–89.
24. Shen, J. C.; Zhang, J.; Letaief, K. B. Downlink User Capacity of Massive MIMO Under Pilot Contamination. *IEEE Transactions on Wireless Communications* **2015**, *14* (6), 3183–3193.
25. Shen, J.; Suyama, S.; Obara, T.; Okumura, Y. Requirements of Power Amplifier on Super High Bit Rate Massive MIMO OFDM Transmission Using Higher Frequency Bands. *Globecom Workshops (GC Wkshps)*, 2014, 433–437.
26. Love, D. J.; Choi, J. Y.; Bidigare, P. A Closed-loop Training Approach for Massive MIMO Beamforming Systems. In *Information Sciences and Systems (CISS), 47th Annual Conference*, 2013; pp 1–5.
27. Kudo, R.; Armour, S.; McGeehan, J. P.; Mizoguchi, M. A Channel State Information Feedback Method for Massive MIMO-OFDM. *J. Commun. Netw.* **2013**, *15* (4), 352–361.

28. Prabhu, H.; Edfors, O.; Rodrigues, J.; Liu, L.; Rusek, F. A Low-complex Peak-to-average Power Reduction Scheme for OFDM Based Massive MIMO Systems. In *Communications, Control and Signal Processing (ISCCSP), 2014 IEEE6th International Symposium on*, May 2014; pp 114–117.
29. Nam, J.; Ahn, J. Y.; Adhikary, A.; Caire, G. Joint Spatial Division and Multiplexing: Realizing Massive MIMO Gains with Limited Channel State Information. In *IEEE 46th Annual Conference on Information Sciences and Systems (CISS)*, March 2012; pp 1–6.
30. Fang, X.; Fang, S.; Ying, N.; Cao, H.; Liu, C. The Performance of Massive MIMO Systems Under Correlated Channel. *19th IEEE International Conference Networks (ICON)*, 2013, 1–4.
31. Li, Y.; Ji, X.; Liang, D.; Li, Y. Dynamic Beamforming for Three-dimensional MIMO Technique in LTE-advanced Networks. *Int. J. Antennas Propagation* **2013**, 2013, 1–8.

CHAPTER 13

Deep Learning in Agriculture as a Computer Vision System

M. SENTHAMIL SELVI*, K. DEEPA, N. SARANYA, and S. JANSI RANI

Sri Ramakrishna Engineering College, Coimbatore 600022, India

Corresponding author. E-mail: senthamilselvi@srec.ac.in

ABSTRACT

In recent days, digital information exchange through social media has been gaining more fame. Sharing of photos and video were increasing day by day. Increase in this information boost, the research in the area of computer vision. Major digital giants like Google, Amazon, Facebook, Microsoft, Apple, etc, are using deep learning techniques in many of their applications. Google's Vision API, Face API, Amazon's Rekognition, AmazonGo are some of the examples where computer vision is used. Computer vision has been using in the fields like agriculture, manufacturing, healthcare, sports tracking, and autonomous vehicle.

13.1   INTRODUCTION

Due to the introduction of deep learning, computer vision-based applications have been gaining popularity. In traditional machine learning algorithm, the features need be extracted and can be used for further processing.[7] Feature engineering plays a major role in machine learning-based applications. Basically, computer vision- or machine vision-based applications consider image as input data. Extracting features from image input is yet another challenging task. It involves image preprocessing, edge detection, segmentation, etc. Introduction of deep learning makes image feature extraction as a single step process. Deep learning has been used in the following machine vision-based applications.[8]

They are:

- Image classification
- Object detection
- Object segmentation
- Image style transfer
- Image colorization
- Image reconstruction
- Image super-Resolution
- Image synthesis

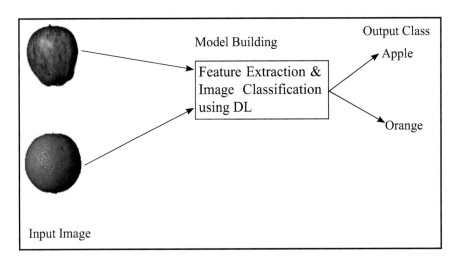

13.2   DEEP LEARNING IN AGRICULTURE

Computer vision-based applications used in agriculture are identification of weeds, land cover classification, plant recognition, fruits counting, ripeness prediction, Crop Monitoring Systems, crop type classification, animal identification and health monitoring, greenhouse climate controller, etc.[2,10] These applications get accelerated due to deep learning and also because of availability of cheap computational resources. Introduction of Precision farming in agriculture makes computer vision-based application a right choice to monitor the farm.

13.2.1   WEED IDENTIFICATION

Identification of weeds in agriculture farms help to spray the herbicides directly on the weeds instead of on the crop. It may be used in conjunction with smart sprayers to facilitate precision herbicide application.[4] Deep learning in computer vision-based application helps the farmers to detect and remove weeds using suitable herbicides.[1,5] This helps to increase the agricultural yield. Due to commercialization of Unmanned Aerial Vehicle (UAV),[11] computer vision-based applications are gaining more importance. Nowadays, drones are equipped with advanced technologies like navigation system, propulsion system, multi rotor, image processing softwares, etc. Drones used in agriculture field helps to monitor crop health, identifying weeds and treating it, seed planting, soil analysis, crop mapping and surveying.[10]

13.2.2   LAND COVER CLASSIFICATION

Land cover refers to physical surface of the Earth's area using which vegetation of the geographical area can be defined.[12] This is useful to analyze the economic growth of any country. Satellite images of the geographic areas help to identify the land cover information which is useful in monitoring agriculture growth, land planning, disaster management, etc. Use of traditional machine learning approaches involves various preprocessing task for feature extraction. Improvement in deep learning helps to simply the work to some extent. Transfer learning techniques were used widely in land cover classification process.

13.2.3   PLANT RECOGNITION

Manual recognition of plant species is more challenging task which requires to identify shape, color, number of petals, existence of thorns or hairs, etc. It is more time consuming process. Deep learning in plant recognition helps the common people to identify herbal plants in their geographic area. This is also useful for the botanist to identify the plant species in very less time without more number of tests.

13.2.4   CROP MONITORING

To improve yield in agriculture field, the crop growth should be monitored consistently. Computer vision algorithms along with intelligent systems can be used for managing and to increase agriculture productivity. Various crop monitoring applications involve measuring plant morphology, measuring growth indicator, diagnosing the nitrogen content, determination of growth stage, and flowering, etc.

13.2.5   FRUIT COUNTING AND RIPENESS PREDICTION

Identifying the ripeness stage of the fruit and counting the fruit were considered vital to automate the fruit monitoring and picking process. Classifying the healthy and rotten fruit is important to assess the quality of any fruit. Fruit maturity prediction is useful in food industries. Agro Robots were designed to pick the fruit based on its maturity level.

13.3   ANIMAL IDENTIFICATION AND COUNTING

Conserving ecosystem and its wealth are important for nation's growth. Preserving wild animals and having updated information about the species and its count are another challenging task. Computer vision with deep learning can be used to accomplish this task.[3] This application minimize the manual work and helps to have accurate statistics about the animal species and its count.

Extracting the appropriate features from the image involves various preprocessing task like background subtraction, segmentation, erosion, dilation, etc. This process makes machine learning a time consuming task. Introduction of Convolutional Neural Network (CNN) makes feature extraction simpler. Accuracy of classification using CNN achieves state of the art result. In all the above mentioned applications, image classification acts as the basic task. Here is a simple example for image classification using CNN is explained.

In this example, image classification is done to identify airplane and car images using python in Windows operating system. Python has been installed through anaconda package. Execution of the code is done through Jupyter notebook Interface.

Steps involved in binary Image Classification using python and keras framework:

- Import basic libraries used for image classification like keras, Num PY, cv2
- Split the dataset as training and testing into a separate folder and subfolder
- Apply one hot encoding technique to represent each class of images in array format
- Represent the input images and its corresponding class label as NumPY array
- Construct CNN model with convolution, pooling, and fully connected layer
- Train the model
- Analyze the accuracy and redesign the network if required
- Test the model with sample image.

13.3.1   STEP 1: IMPORT REQUIRED LIBRARIES

Keras framework is used to build the model. NumPY to represent the data in the form of NumPY array. Matplotlib to visualize the data in the form of graph. Timeit to calculate the total time taken to run the model.

```
from keras.models import Model
from keras.models import Sequential
from keras.layers import Input, Dense, Dropout
from keras.layers import Reshape, Flatten
from keras.layers import Conv2D, MaxPooling2D
import numpy as np
from PIL import Image
import os
from matplotlib import pyplot as plt
%matplotlib inline
import cv2
import timeit
```

13.3.2   STEP 2: SPLIT THE DATASET AS TRAINING AND TESTING INTO A SEPARATE FOLDER AND SUB-FOLDER

13.3.3   STEP 3: APPLY ONE HOT ENCODING TECHNIQUE TO REPRESENT EACH CLASS OF IMAGES IN ARRAY FORMAT

```
def training_image_data():
    train_images=[]
    PATH = 'C:\\DL\\TRAINING'
    # for each folder (holding images of airplane and car)
    for directory in os.listdir(PATH):
        # for each image in the folder
        DATA_PATH=PATH + '\\' + directory
        for image in os.listdir(DATA_PATH):
            img = cv2.imread(os.path.join(DATA_PATH,image)) # convert
            to array
            img=cv2.cvtColor(img,cv2.COLOR_BGR2RGB)
            dim=(64,64)
            new_array=cv2.resize(img, dim, interpolation = cv2.INTER_
            AREA)
            new_array=np.array(new_array)
            if directory=="airplane":
                ohl=np.array([1,0])
            elif directory=="car":
```

```
            ohl=np.array([0,1])
          train_images.append([new_array,ohl])
      print(len(train_images))
      return(train_images)
def testing_image_data():
    test_images=[]
    PATH = 'C:\\DL\\TEST'
    # for each folder (holding images of airplane and car)
    for directory in os.listdir(PATH):
        # for each image in the folder
        DATA_PATH=PATH + '\\' + directory
        for image in os.listdir(DATA_PATH):
            img = cv2.imread(os.path.join(DATA_PATH,image)) # convert
            to array
            img = cv2.cvtColor(img,cv2.COLOR_BGR2RGB)
            dim=(64,64)
            new_array=cv2.resize(img, dim, interpolation = cv2.INTER_
            AREA)
            new_array=np.array(new_array)
            #print(len(new_array))
            if directory=="airplane":
            ohl=np.array([1,0])
            elif directory=="car":
                ohl=np.array([0,1])
            test_images.append([new_array,ohl])
    print(len(test_images))
    return(test_images)
```

13.3.4   STEP 4: REPRESENT THE INPUT IMAGES AND ITS CORRESPONDING CLASS LABEL AS NUMPY ARRAY

```
training_images = training_image_data()
testing_images = testing_image_data()
#Reshape the image to size 64X64 of 3 color channels
```

```
tr_img_data = np.array([i[0] for i in training_images]).reshape(-1,64,64,3)
tr_lbl_data = np.array([i[1] for i in training_images])

tst_img_data = np.array([i[0] for i in testing_images]).reshape(-1,64,64,3)
tst_lbl_data = np.array([i[1] for i in testing_images])
```

Output:

1378

317

13.3.5   STEP 5: CONSTRUCT CNN MODEL WITH CONVOLUTION, POOLING, AND FULLY CONNECTED LAYER

```
np.random.seed(12)
model = Sequential()
model.add(Conv2D(32,  kernel_size=(5,5),  activation='relu',strides=2,
input_shape=[64,64,3]))
model.add(Conv2D(16, (5, 5),strides=2, activation='relu'))
model.add(MaxPooling2D( strides=(2, 2)))
model.add(Flatten())
model.add(Dense(32, activation='relu'))
model.add(Dropout(0.25))
model.add(Dense(2, activation='softmax'))
model.summary()
```

Output: Model Summary describes the model and number of parameters to be trained at each layer

Layer	(type)	Output	Shape		Param #
conv2d_13	(Conv2D)	(None, 30,	30,	32)	2432
conv2d_14	(Conv2D)	(None, 13,	13,	16)	12816
max_pooling2d_5	(MaxPooling2	(None, 6,	6,	16)	0
flatten_7	(Flatten)	(None,	576)		0
dense_13	(Dense)	(None,	32)		18464
dropout_7	(Dropout)	(None,	32)		0
dense_14	(Dense)	(None,	2)		66

Total params: 33,778 Trainable params: 33,778 Non-trainable params: 0

13.3.6 STEP 6: TRAIN THE MODEL

```
start = timeit.default_timer()
model.compile(loss=keras.losses.categorical_crossentropy,
        optimizer=keras.optimizers.Adadelta(),
        metrics=['accuracy'])

history=model.fit(tr_img_data, tr_lbl_data,
    batch_size=128,
    epochs=30,
    verbose=1,
    validation_data=(tst_img_data, tst_lbl_data))
end = timeit.default_timer()
print("Time Taken to run the model:",end - start, "seconds")
```

Output:

Train on 1378 samples, validate on 317 samples

Epoch 1/30

1378/1378 [=====================] - 2s 2ms/step - loss: 0.7591 - acc: 0.8338 - val_loss: 0.7002 - val_acc: 0.8612

Epoch 2/30

1378/1378 [=====================] - 1s 995us/step - loss: 0.4133 - acc: 0.8832 - val_loss: 0.3368 - val_acc: 0.8707

Epoch 3/30

1378/1378 [=====================] - 1s 1ms/step - loss: 0.2999 - acc: 0.8984 - val_loss: 0.3580 - val_acc: 0.9148

Epoch 4/30

1378/1378 [=====================] - 1s 988us/step - loss: 0.3076 - acc: 0.9180 - val_loss: 0.2264 - val_acc: 0.9243

Epoch 5/30

1378/1378 [=====================] - 1s 1ms/step - loss: 0.2050 - acc: 0.9398 - val_loss: 0.2661 - val_acc: 0.9464

Epoch 27/30

1378/1378 [======================] - 2s 1ms/step - loss: 0.0103 - acc: 0.9971 - val_loss: 0.2328 - val_acc: 0.9621

Epoch 28/30

1378/1378 [======================] - 2s 1ms/step - loss: 0.0176 - acc: 0.9956 - val_loss: 0.1305 - val_acc: 0.9621

Epoch 29/30

1378/1378 [======================] - 1s 1ms/step - loss: 0.0562 - acc: 0.9862 - val_loss: 0.0744 - val_acc: 0.9842

Epoch 30/30

1378/1378 [======================] - 1s 1ms/step - loss: 0.9284 - acc: 0.8984 - val_loss: 0.2850 - val_acc: 0.9527

Time Taken to run the model: 45.994907499999954 seconds

13.3.7 STEP 7: ANALYZE THE ACCURACY AND REDESIGN THE NETWORK IF REQUIRED

```
score = model.evaluate(tr_img_data, tr_lbl_data, verbose=1)
print('Training loss:', score[0])
print('Training accuracy:', score[1])
```

Output:

1378/1378 [====================] - 0s 308us/step

Training loss: 0.24737423273169737 Training accuracy: 0.9550072568940493

```
score = model.evaluate(tst_img_data, tst_lbl_data, verbose=1)
print('Validation loss:', score[0])
print('Validation accuracy:', score[1])
```

317/317 [======================] - 0s 422us/step

Validation loss: 0.28495828712165167 Validation accuracy: 0.9526813863203729

13.3.8 STEP 8: TEST THE ACCURACY WITH SAMPLE IMAGE DATA

import pylab as plt

plt.imshow(tst_img_data[200].reshape(64,64,3))

plt.show()

Sample Output

Sample image from Kaggle dataset

Source: https://www.kaggle.com/abtabm/multiclassimagedatasetairplanecar

import numpy as np

prediction = model.predict(tst_img_data[200:201])

print('Prediction Score:\n',prediction[0])

thresholded = (prediction>0.5)*1

print('\nThresholded Score:\n',thresholded[0])

print('\nPredicted object is:\n')

if (np.where(thresholded == 1)[1][0]==1):

    print("car")

else:

    print("airplane")

Output:

Prediction Score: [1.0000000e+00 4.7994902e-12]

Thresholded Score: [0 1]

Predicted object is: car

KEYWORDS

- **deep learning**
- **computer vision**
- **image processing**
- **segmentation**
- **neural network**

REFERENCES

1. Kantipudi Karthik, K.; Lai, C.; Min, C. et al. Weed Detection among Crops by Convolutional Neural Networks with Sliding Windows. *Int. Conf. Precision Agriculture* **2018,** *14,* 1–8.
2. Liakos, K.; Busato, P.; Moshou, D. et al. Machine Learning in Agriculture: A Review. *J. Sensors (Switzerland)* **2018,** 18 (8), 1–28.
3. Norouzzadeh, M.; Nguyen, A.; Kosmala, M. et al. Automatically Identifying, Counting, and Describing Wild Animals in Camera-trap Images with Deep Learning. *Proc. Natl. Acad. Sci.USA* **2018,** 115 (25), E5716–E5725.
4. Olsen, A.; Konovalov, D.; Philippa, B. et al. Deep Weeds: A Multiclass Weed Species Image Dataset for Deep Learning. *Scientific Rep.* **2019,** 9 (1), 1–12.
5. Yu, J.; Sharpe, S.; Schumann, A. et al. Deep Learning for Image-based Weed Detection in Turfgrass. *Eur. J. Agron.* **2019,** 104, 78–84.
6. https://www.forbes.com/sites/bernardmarr/2019/04/08/7-amazing-examples-of-computer-and-machine-vision-in-practice/#5b790e421018
7. https://towardsdatascience.com/cnn-application-on-structured-data-automated-feature-extraction-8f2cd28d9a7e
8. https://machinelearningmastery.com/applications-of-deep-learning-for-computer-vision/
9. https://towardsdatascience.com/7-reasons-why-machine-learning-is-a-game-changer-for-agriculture-1753dc56e310
10. https://www.sbnonline.com/article/commercial-drones-are-increasingly-popular-but-there-are-risks/
11. https://uavcoach.com/agricultural-drones/
12. https://www.tandfonline.com/doi/full/10.1080/15481603.2019.1650447

CHAPTER 14

Deep Learning: Healthcare

M. SENTHAMIL SELVI*, K. DEEPA, S. JANSI RANI, and N. SARANYA

Sri Ramakrishna Engineering College, Coimbatore 600022, India

Corresponding author. E-mail: senthamilselvi@srec.ac.in

ABSTRACT

In recent years, deep learning (DL) methods are powerful because of the massive amount of data and computational power. DL methods works based on the following: raw data fed into machine and based on the data it determines the representation for classification and detection. DL methods/ algorithms works rely on multiple layers of representation of the data, holds important input, and suppresses irrelevant variations. The learning in DL can be supervised or unsupervised. Recent advances in machine learning depend on DL methods.

14.1 INTRODUCTION

DL^2 is a form of representation learning; machine is fed with raw data and develops its own representations needed for pattern recognition that is composed of multiple layers of representations. DL gives immense benefit to the healthcare industry because of the increasing proliferation of medical devices and digital record systems.

So, DL has influence the healthcare industry. In that, accurate medical diagnosis and imaging with DL in computer vision. Based on the survey report from Report Linker that the healthcare market with AI is expected to rise from $2.1 billion in 2018 to $36 billion by 2025.[3]

Two important factors that driven the interest toward the DL in healthcare are:

1. Growth of DL techniques and particularly unsupervised techniques
2. Increase and availability of healthcare data

14.2   BENEFITS OF DEEP LEARNING[4]

Various benefits of DL in healthcare industries are:
Some of them are:

- DL learns the relationship based on the past and it used to diagnosis the patients with same symptoms or diseases.
- Whatever the source data available, DL create the model, when you require a risk score upon administration other than discharge.
- DL enables the confidence and approximate allocation of resources by providing accurate and timely risk scores.
- DL gives better performance and outcomes with low cost.
- When the DL algorithms interact with the training data, they become more specific and accurate allowing individuals to gain exceptional insights into care processes, variability, and diagnostics.
- Graphics processing units or GPUs are getting more efficient to energy and are becoming faster in terms of data.
- DL algorithms are getting sophisticated.
- In healthcare industry, moving toward the electronic health records or EHR and other digitization efforts, this make ease to train data for the DL methods.
- DL algorithms identify the pattern with more precise and faster.
- DL can determine diseases or diagnosis like certified professionals.

14.3   USE OF DEEP LEARNING IN ELECTRONIC HEALTH RECORDS[5]

Nowadays, Patient's history/record in hospitals is converted and maintained as an electronic version, named as EHR. It includes all the records like demographics of the patient, problems, progress, vital signs, past medical history, immunizations, laboratory details, and other reports related to the patient. EHR automates access to information and has the potential to streamline and simplify the clinician's workflow. EHR strengthen the relationship between the patients and clinicians and it provides to make better decisions and care.

Example:

- Possibility of reduces the medical error
- Better decision during critical situation without delay of medical process
- Accuracy of clarity of medical record

EHR[3] improves with the help of DL methods and algorithms. DL algorithms help to identify the patterns from EHR data, this in turn helps to identify the risk factor and make conclusion based on the pattern derived from the algorithm.

Two ways to use EHR system data:

1. A static prediction: A static prediction gives likelihood of an event based on a dataset fed into the system and code embeddings from the International Statistical Classification of Diseases and Related Health Problems (ICD).
2. A prediction based on a set of inputs: EHR system driven data are used to make a prediction based on a set of inputs. It is possible to make a prediction with each input or with the entire dataset.

14.4 GENERATIVE ADVERSARIAL NETWORK

EHR data are not appropriate for sometimes like when it is used to diagnose the rare disease or for procedures for unique diseases. To make up the above problem or to solve, doctors and researchers used Generative Adversarial Network (GAN) DL method.

GAN are deep neural net architecture comprised of two nets, pitting one against the other (thus it is "adversarial"). One is generator and other one is discriminator. GANs' potential is vast because they can learn to mimic any distribution of data.[6]

The important data in a given dataset learn by the generator and to deceive the discriminator to think genuine, the generator will generate the new data instances. For authentication, both dataset tested by the discriminator and decide the result as fake or real (real [1] and fake [0]). To produce the better result and quality, the above process repeats by forcing the generator too keep train for learning the data.[3]

14.5   EXAMPLE APPLICATIONS IN HEALTHCARE

To reduce the rate of misdiagnosis and predicting the outcome of procedures, DL techniques use data stored in EHR system to address needed healthcare concerns. By processing large amounts of data from various sources such as medical imaging, ANNs can help physicians/doctors to analyze information and detect multiple conditions like:

- Blood sample analysis
- Tracking glucose level of Diabetic patients
- Detect heart related problems
- Tumor detection using image analysis
- Detecting cancerous cells for diagnosing cancer
- Detecting osteoarthritis from an MRI scan before the damage has begun

Some applications related to radiology are:[7]

- Screening of lung cancer from CT scans
- Automatic tumor characterization and detection
- Analysis of brain images
- Gene expressions profiling in several types of cancers
- All CAD system are trying to migrate into DL technologies
- Segmentation of anatomical objects from medical scans

So, DL helps medical professionals and researchers to discover unseen opportunities in data and helps for the betterment of healthcare industry. The DL method provides better analysis of any diseases and provides help to treat them efficiently and effectively. Thus, results accurate medical decisions by right time. Some more applications are:

14.5.1   DRUG DISCOVERY

Help in discovery of drugs and their development. DL methods analyze the patients' medical history and provide best medicine for them. It depends on the data available like patient symptoms and test records.

14.5.2 MEDICAL IMAGING

Techniques such as CT scans, MRI scans, ECG reports are used to diagnose the diseases such as heart, tumor, etc. Hence, DL techniques provide better results on those medical images.

14.5.3 INSURANCE FRAUD

DL used to analyze the fraud claims. Also with predictive analytics, it can predict the likelihood of future fraud claims. It helps for insurance industry to rise up for discounts and offers for the customers.

14.5.4 ALZHEIMER'S DISEASE

DL techniques identify the disease at the early stage.

14.5.5 GENOME

DL used to understand and analyze the genome and helps patient to get idea about the disease that may affect them in future.

14.6 3 WAYS DEEP LEARNING IS REINVENTING THE HEALTHCARE INDUSTRY

People assigning more importance toward their health. The expectation from the doctors is the highest level of care and services regardless of cost. The ways for better the patient care are:

1. Faster Diagnosis
2. Genomics for personalized medicine
3. Computer-aided diagnosis (CADX)

14.7 PREDICTIVE ANALYTICS IN HEALTHCARE: BENEFITS?[8]

14.7.1 BUSINESS PROCESS: BETTER OPERATIONAL EFFICIENCY

Analytics plays vital role in healthcare industry to enhance the efficiency of business process by analyzing the patient dataset to determine admission and re-admission rates. Also, used to monitor and analyze the hospital staff performance in real-time.

14.7.2   ACCURACY: IDENTIFICATION AND TREATMENT THROUGH PERSONALIZED MEDICINE AND DRUGS

Every person needs special care and attention. New technology intervention helping the doctors to analyze the individual patient based on their records for identification of diseases and treatment. This gives stratification to patient and doctors.

14.7.3   ENHANCEMENT: COHORT TREATMENT

To make better decision and crucial decision, the digitization of patient records is used and makes it possible. Predictive analytics in healthcare includes people studies that use huge volumes of patient data to create profiles of community and other cohort health patterns to create early interventions that aim to reduce the financial and resource load healthcare organizations.

14.8   CHALLENGES AND LIMITATIONS OF DEEP LEARNING IN HEALTHCARE[9]

In compare with traditional machine learning approaches, DL can bring substantial improvements, many researchers and scientists remain sceptical of their use where medical applications are involved. DL methods have not provided complete solutions and many questions remain unanswered. The following are the aspects review where some of the potential issues associated with DL:

1. Recent work by researchers, the entire DL model is often not interpretable when CNN with weight filters to visualize the high level features. So DL methods and approaches used as a black box by most of the researchers. Because it is not possible to explain or interpret why methods give better result or ability of modification to handle the misclassification problems.
2. Large sets of training data are required to train a reliable and effective model to express a new concept. In recent years, healthcare

industries transform all paper records to electronic records. The dataset with less number samples and less parameters of the Deep Neural network (DNN) produces better result. So DNN training is overfitting for small datasets. In this case, the model learns and memorizes the training set and works well for testing dataset but gives error or less accuracy to generalize the new samples that are not already learned. The accuracy of the learning samples is high when compared to new data samples (very less accuracy). To avoid the above overfitting problem and improve generalization, regularization method such as the dropout is usually exploited during training.

3. In many application, data are raw data; the handling of raw data is the next important aspect in DNN. However, the raw data cannot directly used as input for DNN. Before training the data, raw data can be preprocessed or changes in domain are required. The arrangement of hyper parameters setup for controlling architecture of a DNN, which includes size and the number of filter in a CNN, or its depth, is still a blind exploration process that usually requires accurate validation. For efficient classification model, the following are important:

• Find and implement the perfect preprocessing method for the data
• Choose and setup the optimal set of hyperparameters

4. The final aspect most of the DNN fooled easily by adjusting the parameter in the data. If any small changes into the input sample such as unnoticeable noise in an image produces the high misclassification rate (i.e., samples are misclassified). In logistic regression, intentionally increase or decrease the value of feature influences the misclassification and its rate. Likewise, for decision tress, a single binary feature can be used to direct a sample along the wrong partition by simply switching it at the final layer. Therefore in general, any machine learning models are susceptible to such manipulations.

14.9 CHALLENGES AND OPPORTUNITIES[10]

The following are also the issues facing the clinical applications of DL to healthcare. In particular,

Data volume: Comparing with traditional model DL requires more data. In general, there are no proper guidelines about the training documents, but based on the rule of thumb is to have at least 10 number of samples as parameters in the network. In healthcare, the available data are less when compare with the population in the world.

Data quality: Healthcare data are heterogeneous, ambiguous, noisy, and incomplete when compare with other data. Training a good DL model with such huge and variety datasets is challenging and needs to consider several issues, such as data sparsity, redundancy, and missing values.

Temporality: The model developed and proposed are able to handle the static input, it does not handle the time related factor because disease and its types are evolving and changing over time. So the developing the DL must handle temporal healthcare data.

Domain complexity: In some cases, there is less knowledge on progress of the diseases. Sometimes the number of cases and patients reported are less, cannot get relevant data from such cases.

Interpretability: Although DL models have been successful in a few application domains, they are often treated as black boxes. Though this might not be a problem in other more domains such as image annotation (because the end user can objectively validate the tags assigned to the images), in healthcare, not only the quantitative algorithmic performance is important, but also the reason why the algorithms works is relevant. Such model interpretability (i.e., providing which phenotypes are driving the predictions) is influential the medical professionals about the actions recommended from the predictive system (e.g., prescription of a specific medication and potential high risk of developing a certain disease).

14.10   RESEARCH OPPORTUNITY

All the above challenges give research opportunity in healthcare domain. With this, the following directions would be future of DL in healthcare domain.

Feature enrichment: Patient data in the healthcare is limited, for future improvement of DL system; the system should capture more features in all ways and find the novel method to join them for process. The data sources to generate those features need to include, but not to be limited to, EHRs, social media, wearable devices, environments, surveys, online communities, genome profiles, and so on. The integration of all scattered and heterogeneous data and how to use them and implement in a DL model would be a significant and demanding research topic.

Federated inference: DL system with security is difficult (in terms of handling sensitive information about patient).

Model privacy: For scaling the DL, privacy is an important concern. The deployment of tools should handle risks and must implement privacy standards.

Incorporating expert knowledge: Due to quality and limited amount of medical data, important to incorporate the expert knowledge into the methods to move in right direction and get better performance.

Temporal modeling: Time is critical factor for healthcare data. The DL method must consider the temporal data.

Interpretable modeling: There is the equal importance for the system performance and interpretability in healthcare problems. The system with less and cannot understand are not like to adopt by clinicians. DL models are popular because of their better performance. Also how to explain the results obtained by these models and how to make them more understandable is of key importance toward the development of trustable and reliable systems.

14.11   CURRENT AND FUTURE DIRECTIONS IN HEALTHCARE AND MEDICAL DOMAINS[4]

- Different kinds of data are emerging in fast world. Gaining knowledge from the heterogeneous data and high dimensional are challenges in healthcare industry. DL solves the issues of imaging, sensor text, sensor data, and electronic records.
- Usage of EHR promises to move forward clinical research and better inform the decision-making skills clinically. Modern EHR can avoid the practice of predictive modeling by summarizing and representation. Patients who achieved results based on EHR data and an alternative feature learning strategy are being unbeaten. DL framework arguments decision systems in the clinical environment.
- DL, acute peripheral arterial thrombosis, and embolism along with acute coronary thrombosis can be treated.
- Application of this method (method: Program by learning itself), there is a further assessment and validation to medical imaging.

- The other areas of DL in healthcare are classification of medical images, segmentation, registration, and other tasks. Also, used in other areas such as retinal, digital pathology, pulmonary, neural issues, etc. In data analysis, DL is a progressively developing trend and has also been named one of the revolution technologies in recent years.
- DL learned model useful in many applications and in turn methods gained knowledge from health care data.
- Due to the large volume and veracity of data in healthcare, it is in the primary focus of DL and machine learning researchers and experts of the industries.
- Large datasets from clinical management systems increase the healthcare sectors. This gives a strength and option for any application of DL approaches on healthcare datasets, which may be sparse.

14.12   CONCLUSION

DL can provide the path toward the development of the healthcare industry. The advancements in DL help to solve the problems and provide effective ways to handle complex data structures.[4]

KEYWORDS

- **deep learning**
- **computer vision**
- **health care**
- **segmentation**
- **neural network**

REFERENCES

1. http://www.jogh.org/documents/issue201802/jogh-08-020303.pdf
2. https://www.researchgate.net/publication/330203264_A_guide_to_deep_learning_in_healthcare

3. https://missinglink.ai/guides/deep-learning-healthcare/deep-learning-healthcare/
4. https://technostacks.com/blog/deep-learning-in-healthcare/
5. https://www.cms.gov/Medicare/E-Health/EHealthRecords/index
6. https://skymind.ai/wiki/generative-adversarial-network-gan
7. https://www.quora.com/What-are-the-major-applications-of-deep-learning-in-medicine
8. https://www.businesswire.com/news/home/20191009005448/en/Pros-Cons-Predictive-Analytics-Healthcare-Quantzig%E2%80%99s-Latest
9. Ravi, D.; Wong, C.; Deligianni, F.; Berthelot, M.; Andreu-Perez, J.; Lo, B.; Yang. Deep Learning for Health Informatics. *IEEE J. Biomed. Health Info.* **2017,** *21* (1), 4–21.
10. Miotto, R.; Wang, F.; Wang, S.; Jiang, X.; Dudley, J. T. Deep Learning for Healthcare: Review, Opportunities and Challenges. *Brief. Bioinfo.* **2017,** *19* (6) 1236–1246; http://dudleylab.org/wp-content/uploads/2017/05/Deep-learning-for-healthcare-review-opportunities-and-challenges.pdf

CHAPTER 15

Infrastructure Health Monitoring Using Signal Processing Based on an Industry 4.0 System

SAMEER PATEL[1*], AJAY KUMAR VYAS[2], and KAMAL KANT HIRAN[3]

[1]Department of Civil and Infrastructure Engineering, Adani Institute of Infrastructure Engineering, Ahmedabad, India

[2]Department of Electrical Engineering, Adani Institute of Infrastructure Engineering, Ahmedabad, India

[3]Research Fellow, Aalborg University, Copenhagen, Denmark

[*]Corresponding author. E-mail: sameer.patel@aiim.ac.in

ABSTRACT

Vibration signal processing is one of the most efficient techniques for the collection of data using the sensors. Nowadays, development in wireless sensors and their network technologies gives a better platform to evaluate condition assessment of bridge structures remotely based on cloud-based data collection using Industry 4.0 concept. This technique can be used for continuous monitoring of structural behavior, especially for bridges. In this chapter, two different testing techniques forced vibration and ambient vibration are applied on the test bridge for evaluation of structural condition using subspace system identification and least-squares complex frequency approach. Comparative study of forced vibration testing and ambient vibration tests shows that only 3–4% variation and natural frequency of the bridge varies from 5.50 to 5.62 Hz. It is concluded that wireless sensor networks based on Internet of Things (IoT) give satisfactory result without any data loss during cloud computing. It is also observed that the ambient vibration test is more suitable than the force vibration test as it does not require traffic shutdown while performing the field test. Therefore, the evaluated technique can also use on various smart infrastructure projects, where dynamic behavior of the structure is essential to evaluate based on the continuous monitoring.

15.1 INTRODUCTION

Infrastructure development is becoming important for the nation-building of the country. Infrastructure comprises ornamental the superiority of a life of the urban population and growing the city's affordability, bridges, dams, roads, urban infrastructure development, refining logistics, transportation, power, energy, agri-business, information and communication technology. Under the Union Budget 2018–19, US \$92.22 billion was allocated to the sector. As a developing country, India has a requirement of investment worth 50 trillion (US \$777.73 billion) in infrastructure by 2022 to have sustainable development in the country.[1] Along with this, the maintenance cost of existing infrastructure also becomes a key financial aspect for the government authority to evaluate safety and serviceability criteria of existing infrastructure usage and up-gradation of existing infrastructure projects.

Deterioration of civil infrastructure systems such as bridges, buildings, and pipelines, is a complex problem due to ageing, distress, increasing loading conditions, ambient environmental condition, etc. Therefore, periodic and continuous monitoring of structures becomes a prime interest of researchers to evaluate or assess structural condition with non-destructive evaluation techniques since the past few decades.[2]

In health monitoring of bridges, the most common technique used to evaluate the present condition of bridges is visual inspection, reliability of subjectively collected data highly depends on the skills and experiences of the evaluator. Recently, the development of sensor networks and systems could be helpful to assist the authorities to predict the exact condition of the bridges.[3] A vibration-based health monitoring of structures becoming popular as it is the most efficient technique to investigate structural strength and durability parameters of the civil engineering structures without damaging structures like existing conventional techniques. In vibration-based monitoring, identification of changes in the dynamic characteristics of the structure obtained from recorded structural vibration signals data induced by ambient and artificial loading sources.[4] It is noticed that bridge damages lead to change bridge properties such as mass, stiffness, and damping of the structure due to that dynamic characteristics are getting affected, such as natural frequencies and mode shapes. Therefore, after investigating the dynamic characteristics, one can predict the intensity of damage and service life of the specific structures.[5] Since the last four to five decades, rigorous studies on monitoring of bridges as well as other structures are carried out with large numbers of experiments. Different theories are also developed during these years. Along with these developments in digital computers, analytical software packages, data acquisition systems are noticeably observed. Recently, all systems operate online via a high-speed wireless internet network, allowing real-time data transmission,[6] which gives a better platform to evaluate condition assessment of bridge structures remotely based on cloud-based data collection using Industrial 4.0 concept.

In this chapter, we analyzed the vibration-based health monitoring system for slab deck type of concrete bridges. The sets of vibration measurements are conducted at a test bridge and comparative study carried out based on the ambient and forced vibration technique using LabVIEW 12.0 and ModalVIEW. The investigation is reliable for evaluation of the structural condition of the bridge and it does not require the traffic shutdown.

15.2   VIBRATION TESTING

The amplitude and the frequency of the vibration of the bridge can be controlled by the excitation applied and the response of the bridge to a particular excitation. There are various conditions which can affect the response of the bridge excitation for vibration testing like a method of excitation, the position of the exciting force, or motion in addition to dynamic characteristics of the bridge such as its inherent damping level and the natural frequencies. The vibration testing method is classified on the bases of the application of excitation force as forced vibration testing (FVT)[7] and ambient vibration testing (AVT).[8] The health of the bridge is can be monitor by these vibration testing methods with precisely. Wireless smart sensor networks are having advantages over wired sensor network as they are less expensive and easier to fix it on any remote location on the bridges.[9] The statistical analysis of system identification results determines the actual modal parameters of bridge such as mode shapes, modal frequencies, and damping properties.[10]

15.3   EXPERIMENT SETUP

The test bridge has a solid slab-deck type of superstructure with 11 spans of
12.0 m each as shown in Figure 15.1. It is a clear roadway with a width of
7.50 m that carries a unidirectional traffic flow on it. The depth of the slab-
deck was 1.025 m. A detail dimension of the bridge is shown in Figure 15.1.
The span of the bridge was instrumented with eight accelerometers above the
deck and besides the railing for performing AVT as shown in Figure 15.2.

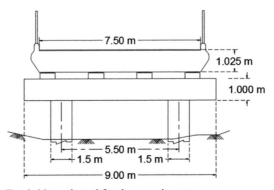

FIGURE 15.1   Test bridge selected for the experiment.

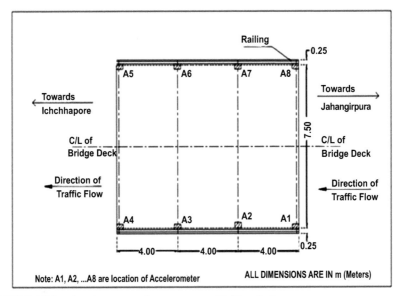

FIGURE 15.2   Accelerometers layout on tested bridge deck.

In addition to ambient vibration testing, free vibration test was also performed using construction vehicles for the experimental study. The dimension of the tested truck is shown in Figure 15.3. The test truck had three axles and was weighed before the test to determine the axle loads.

We performed a free vibration test with a known weight of vehicles (i.e., RMC Miller) under empty, half, and fully loaded conditions (Fig. 15.3) with different moving speed. The acceleration time histories were recorded during the moving truckload at 20, 30, and 40 kmph speeds.

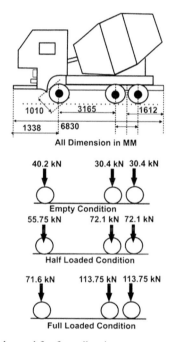

FIGURE 15.3 Test vehicle used for free vibration test on test-bridge.

15.4 RESULTS

In the ambient vibration test, vibrational signals were captured through accelerometers at predefined locations, as mentioned in the previous section. All signals were recorded in the form of acceleration time histories. These time-domain signals were further used for post-processing to perform modal analysis, through which modal parameters were determined. Acceleration time history observed throughout an AVT of the test bridge as shown in Figure 15.4.

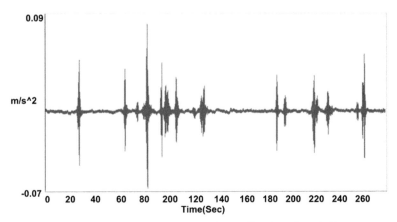

FIGURE 15.4   Acceleration time history of ambient vibration of test bridge.

The amplitude of the acceleration signal generated from the accelerometers is averaged to calculate the global pointer or the vibration levels
each hour. The distribution of the averaged acceleration signal amplitude
is shown in Figure 15.5. The distribution is appearances that the variation
of the amplitude is day time from 09:00 am (i.e., anti-meridian) to 12:00
pm (i.e., post meridiem) and during the evening from 04:00 to 07:00 pm at
0.095–0.167 and 0.13–0.188 m/s², respectively.

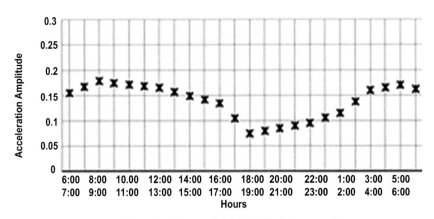

FIGURE 15.5   Averaged acceleration amplitudes (m/s²) during tested hours.

It was preferable to have two complete vibration tests (i.e., ambient and
forced vibration test) of the same bridge under similar conditions. Least-
squares complex frequency (LSCF) and subspace system identification

method are adopted for data post-processing. In the subspace system identification (SSI) method, the stochastic subspace model is designed to calculate the model frequency and the stabilization diagram is generated by identifying state-space models for orders n = 80. The correlation can be defined as per the vibration signal frequencies; it is 1% for frequencies and 5% for both damping, and mode shape. The stabilization diagram obtained after performing modal analysis using the SSI technique shown in Figure 15.6.

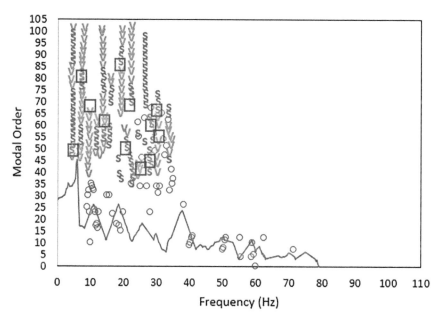

FIGURE 15.6 Modal frequencies identification of test bridge by SSI method.

In the LSCF method, the stabilization diagram was constructed by defining the model orders, n = 80. The criteria for correlation are the same as SSI method 1. The stabilization diagram obtained after performing modal analysis using the LSCF technique is shown in Figure 15.7. For the corresponding modes, modal frequencies were evaluated by operation analysis method in AVT, with corresponding values for tests as given in Table 15.1. The uncertainty arises from the variability in ambient conditions that affect the measured frequencies. It was seen that there are eight possible modes of frequency below 35 Hz. It was observed that the fundamental frequency was 5.611 and 5.617 Hz (by calculating with the SSI and LSCF methods, respectively). These mode shapes corresponded to the bending, torsional, or combined effect. Statistical analysis of frequency results indicated that the maximum relative difference of frequency estimates of the bridge by both the methods may vary from 1 to 4% and the damping ratio varies from 0.5 to 3%. The reference frequency for the ambient vibration test is taken 5.617 Hz.

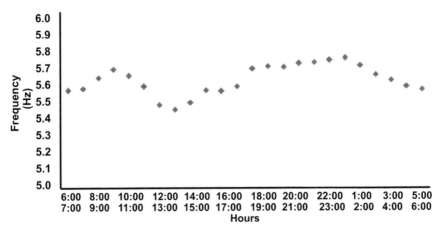

FIGURE 15.7   Modal frequencies identification of test bridge by LSCF method.

For the corresponding modes, the modal frequencies are found by the modal analysis method in FVT, with corresponding values for tests as given in Table 15.2. A total of eight possible modes shapes are extracted below 35 Hz. These mode shapes corresponded to the bending, torsional, or combined. It is observed that the fundamental frequency was 5.602 Hz and that it varied from 5.587 to 5.602 Hz under different loading conditions as well as speed of the test vehicle.

TABLE 15.1   Results of Mode and Modal Frequencies and Damping Ratio from AVT of Bridge.

Mode	Frequency (Hz)		Damping ratio (%)	
	SSI method	**LSCF method**	**SSI method**	**LSCF method**
1	5.611	5.617	0.205	0.21
2	9.514	9.615	0.199	0.201
3	14.591	14.389	0.181	0.175
4	19.501	19.647	0.146	0.144
5	24.152	24.162	0.146	0.138
6	27.288	27.261	0.122	0.112
7	29.590	29.485	0.086	0.081
8	31.824	31.683	0.064	0.06

TABLE 15.2 Results of Mode and Modal Frequencies from FVT of Test Bridge.

Mode	Truck loading with Speed 20 kmph			Truck loading with Speed 30 kmph			Truck loading with Speed 40 kmph		
	Empty	**Half loaded**	**Full loaded**	**Empty**	**Half loaded**	**Full loaded**	**Empty**	**Half loaded**	**Full loaded**
1	5.587	5.589	5.591	5.590	5.593	5.595	5.594	5.598	5.602
2	9.467	9.470	9.475	9.471	9.476	9.478	9.477	9.481	9.485
3	14.512	14.521	14.527	14.522	14.527	14.530	14.529	14.533	14.538
4	19.465	19.472	19.477	19.474	19.479	19.482	19.481	19.485	19.489
5	24.108	24.111	24.117	24.112	24.119	24.121	24.120	24.124	24.129
6	27.254	27.258	27.263	27.260	27.264	27.268	27.266	27.270	27.275
7	29.496	29.499	29.501	29.498	29.504	29.509	29.506	29.509	29.513
8	31.787	31.791	31.794	31.790	31.796	31.801	31.798	31.802	31.808

The experiment results show that the test vehicle (RMC Transit Mixer) moving less than 40 kmph, the bridge frequency could be easily identified. However, as the test vehicle speed increases, the bridge frequency blurs due to the involvement of high-frequency components resulting from the cart structure and pavement roughness. The existence of ongoing traffic is considered beneficial since it tends to intensify the truck response.

Hourly identification of the natural frequency of the bridge is shown in Figure 15.8 and it is observed that the hourly based natural frequency of the bridge varies from 5.50 to 5.62 Hz.

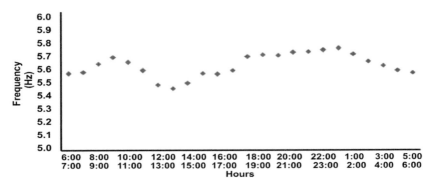

FIGURE 15.8   Hourly based identification of natural frequency of test bridge.

15.5   CONCLUSION

The single span of the bridge would have a total of eight (8) sensors installed. These accelerometers (sensors) in one span are called a cluster. The sensor and its network configuration provide each sensor node to connect with wireless data acquisition network, which has the ability to connect with the internet in order to transmit data to the central server, ready for remote access by the authorized users. The SHM System designed would measures the actual acceleration data on the bridge in real-time, and compare with data to thresholds set through the structural analysis of the bridge as an indicative alarm.

The ambient vibration tests carried out using traffic-induced vibration for slab-deck type of concrete bridge superstructures. The two different vibration analysis approaches such as SSI and LSCF are applied for modal parameters extracted. It is concluded that SSI methods are more accurate and easy to extract modal parameters directly from time-domain signals captured with a wireless sensor network.

Forced vibration test is also conducted with known weight and speed of test vehicles. It is observed that low speed (less than 20 km) of test vehicle does not create a large dynamic impact on the bridge so that the amplitude of vibration remains low. The higher speed of the test vehicle produces sufficient amplitude of vibration but disturbance or noise in signals are also recorded vibration signals. Before performing modal analysis, it is necessary to remove unwanted frequencies from the vibration signals by using the anti-aliasing filter and bandpass filters. So that bridge natural frequency can be identified with higher accuracy. The concluding remark of the experiment shows that only 3–4% variation in results is observed. It is concluded that ambient vibration test is more suitable than the force vibration test as it does not require traffic shutdown while performing the field test.

ACKNOWLEDGMENTS

Authors would like the express acknowledgement to Dr G R Vesmawala, Department of Applied Mechanics, Sardar Vallabhbhai National Institute of Technology Surat for his guidance and support.

KEYWORDS

- **ambient vibration test**
- **bridge health monitoring**
- **force vibration test**
- **Industry 4.0**
- **wireless sensor network**

REFERENCES

1. Indian Infrastructure Sector in India Industry Report, November 2018; www.ibef.org
2. Yu, Y.; Xie, H.; Ou, J. Vibration Monitoring Using Wireless Sensor Networks on Dongying Huanghe River Bridge. *Earth and Space 2010: Engineering, Science, Construction, and Operations in Challenging Environments, ASCE*, 2010.
3. Hendricks, M. D.; Meyer, M. A.; Gharaibeh, N. G.; Van Zandt, S.; Masterson, J.; Cooper Jr, J. T.; Horney, J.A.; Berke, P. The Development of a Participatory Assessment Technique for Infrastructure: Neighborhood-level Monitoring Towards Sustainable Infrastructure Systems. *Sustain. Cities Soc.* **2018**, *38*, 265–274.
4. Shit, R. C.; Sharma, S.; Puthal, D.; Zomaya, A. Y. Location of Things (LoT): A Review and Taxonomy of Sensors Localization in IoT Infrastructure. *IEEE Commun. Surveys Tutorials* **2018**, *20* (3), 2028–2061.
5. Tanwar, S.; Tyagi, S.; Kumar, S. The Role of Internet of Things and Smart Grid for the Development of a Smart City. *Intelligent Communication and Computational Technologies. Lecture Notes in Networks and Systems*, Singapore; 2018, pp 23–33.

6. Fraser, M. et al. Sensor Network for Structural Health Monitoring of a Highway Bridge. *J. Comput. Civ. Eng., ASCE* January/February **2010,** 11–24.

7. Star, L. M.; Tileylioglu, S.; Givens, M. J.; Mylonakis, G.; Stewart, J. P. Evaluation of Soil-structure Interaction Effects from System Identification of Structures Subject to Forced Vibration Tests. *Soil Dyn. Earthquake Eng.* **2019,** *116,* 747–760.

8. Roselli, I.; Malena, M.; Marialuisa, M.; Cavalagli, N.; Gioffrè, M.; Canio, G. D. Health Assessment and Ambient Vibration Testing of the "Ponte delle Torri" of Spoleto During the 2016–2017 Central Italy Seismic Sequence. *J. Civ. Struct. Health Monitor.* **2018,** *8,* 199–216.

9. Asadollahin, P.; Jian, L. Statistical Analysis of Modal Properties of a Cable-Stayed Bridge Through Long-Term Wireless Structural Health Monitoring. *J. Bridge Eng., ASCE* **2017,** *22* (9), 1–15.

10. Xu, Y.; Brownjohn, J. M.; Hester, D. Enhanced Sparse Component Analysis for Operational Modal Identification of Real-life Bridge Structures. *Mech. Syst. Sign. Process.* **2019,** *116,* 585–605.

Index